T0215273

MEDIATING
THE HUMAN BODY

Technology, Communication,
and Fashion

MEDIATING THE HUMAN BODY

Technology, Communication, and Fashion

Edited by

LEOPOLDINA FORTUNATI
University of Udine, Italy

JAMES E. KATZ
Rutgers, The State University of New Jersey, USA

RAIMONDA RICCINI
Istituto Universitario di Architettura di Venezia, Italy

Routledge
Taylor & Francis Group

NEW YORK AND LONDON

The Human Body Between Technologies, Communication, and Fashion conference was sponsored by:

First published 2003 by Lawrence Erlbaum Associates, Inc.

Published 2010 by Routledge
4 Park Square, Milton Park, Abingdon, Oxon OX14 4RN
605 Third Avenue, New York, NY 10017

Routledge is an imprint of the Taylor & Francis Group, an informa business

Cover design by Kathryn Houghtaling Lacey

Cover image by Stefano Mandato, Chiara Rolfini, and Umberto Tolino

Library of Congress Cataloging-in-Publication Data

Mediating the human body : technology, communication, and fashion / edited by Leopoldina Fortunati, James Katz, Raimonda Riccini.
    p. cm.
Based on a conference held Jan. 11–12, 2001, during the Triennale di Milano.

    Includes bibliographical references and index.
ISBN 0-8058-4480-5 (cloth : alk. paper)
ISBN 0-8058-4481-3 (pbk. : alk. paper)
1. Body, Human—Social aspects. 2. Human beings—Effect of technological innovations on. 3. Human figure in art. 4. Human-computer interaction. 5. Communication and technology. 6. Mass media and technology. 7. Technology and civilization. 8. Advertising—Fashion. I. Fortunati, Leopoldina. II. Katz, James Everett. III. Riccini, Raimonda.
GN298. M44 2003
306.4—dc21
                                2002035395
                                   CIP

ISBN 13: 978-0-8058-4481-8 (pbk)

Vincendo me col lume d'un sorriso
ella mi disse: "Volgiti ed ascolta,
che non pur nei miei occhi è paradiso"

19–21
Canto XVIII
Il Paradiso
La Divina Commedia di Dante Alighieri

Winning me with her radiant smile,
said she: "Turn thee around and listen:
Paradise is not in my eyes alone."
Editors' translation

Contents

Acknowledgments

The editors thank the Department of External Relations of Omnitel–Vodafone, Anna Maria Grossi of the Fashion and Events Department of Milan City Council, the Aventis Foundation (Italy), and Diego Mocellini of Bli.io for their generous support of the international conference "Il corpo umano tra tecnologie, comunicazione e moda / The Human Body Between Technologies, Communication and Fashion," and this publication.

Alessandra Vecchio of Franco Angeli Editore was most understanding and constructive, whereas Jo Falinski has given significant assistance with many of the translations. Liz Paton's skillful copyediting has helped us avoid some pitfalls and everywhere contributed to the manuscript's lucidity.

Professors William Dutton, Kenneth J. Gergen, Irving Louis Horowitz, Tomás Maldonado, Robert K. Merton, and Josh Meyrowitz helped by pointing us to some unexamined issues concerning the body, technology, and communication. Professors Arturo Dell'Acqua Bellavitis, Ronald Day, Claire McInerney, and Ronald E. Rice have been true colleagues, selflessly contributing ideas and assistance throughout this intellectual voyage.

Linda Bathgate of Lawrence Erlbaum Associates, a highly talented Renaissance woman, has been a delight to work with, and her thoughtful guidance is much appreciated. Karin Wittig Bates helped us expeditiously surmount production hurdles. Also, we gratefully acknowledge our Production Editor, Marianna Vertullo.

To all these kind and supportive individuals, we offer our heartfelt gratitude.

—L. F.
—J. E. K.
—R. R.
May 2002

Foreword

Pierfrancesco Gamba
Alderman, Fashion and Events Department, Milan City Council

The chapters of this book are based on the international conference "Il corpo umano tra tecnologie, comunicazione e moda / The Human Body between Technologies, Communication and Fashion," January 11–12, 2001. The conference was held at one of Italy's most renowned cultural institutions, the Triennale di Milano, which as a center of art and culture, and both a museum and a center of creativity, was an ideal location. The Fashion and Events Department of Milan's City Council was a major sponsor of this conference, although many other important institutions also joined in sponsorship. The Department undertook sponsorship to help foster progress in an area of concern for the Department, and indeed for the leadership of the city of Milan, namely advancing analysis of trends in fashion, design, and art, as well as fostering dialogue and interaction among the communities of experts concerned with these issues. As Alderman for this Department, I am pleased to have this opportunity to offer a perspective on the importance of the conference.

The conference had a noble goal: to explore the effects of the new technologies on human beings. The specific focus was the intersection of new communication technologies and the body. Conference sponsorship is an important part of the Fashion and Events Department's continuing promotion of meaningful interaction between those concerned with arranging fashion events in Milan, the companies involved with technological operations that produce material for the fashion industry, and scholars working in the new languages of communication. The Fashion and Events Department believes that occasions such as this allow the leaders of these systems to meet and discuss issues of fundamental and mutual importance. These include the cultural and economic future of the Milan region and the advancement of the quality of life throughout the industrial world. The conference offered the opportunity to learn from each other's experience and to create a novel, powerful synthesis between the cultural and operational spheres.

It is significant that it was the city of Milan that hosted a conference embodying the state of the art in these matters. The conference offered an authoritative and privileged observation point from which to study lifestyles and new trends. This conference also reflected the importance of culture in Milan. The Fashion and Events Department has often given space and encouragement to similar occasions in the past, and believes the entire city benefits from the accompanying intellectual ferment and creativity. As a result, investments and resources are attracted to Milan within the framework of global competition.

In the Age of the Information Economy, the style of garments is changing. Different kinds of communication technology are emerging from the cybernetic present, and new horizons are opening up before firms that once clothed the denizens of the Old Economy.

Our commitment has been to advance lifestyle and intellectual initiatives that have confirmed Milan as a leader in the field of fashion and design. The Department has become a point of reference for those seeking to enhance the creative richness of the human experience. In turn, these activities encourage and support specific competitive activities that, in the long run, benefit all. Milan is a melting pot of ideas, unstinting in its encouragement of new projects and initiatives, and untiring in its pursuit of putting the best of them into practice. It is this balanced commitment to the intellectual and material well-being of its citizenry that makes Milan an exciting and desirable place to live, work, and visit.

Contributors

Mark Aakhus is an Assistant Professor of Communication at Rutgers University School of Communication, Information and Library Studies, USA. He coedited, with James Katz, *Perpetual Contact: Mobile Communication, Private Talk, Public Performance* (Cambridge University Press, 2002).

Patrizia Calefato teaches sociolinguistics and philosophy of language at the University of Bari, Italy. Her interests are theory of language, the socio-semiotics of culture and communication, the semiotics of fashion, and feminist studies. She is vice president of the Italian Association of Semiotic Studies and of the Italian Society of Women Literati. She sits on the International Advisory Board of the review *Fashion Theory*. Her latest publications include: *Moda, corpo, mito* (Castelvecchi, 1999) and *Lingua e discorso sociale* (Graphis, 2001).

Massimo Castellani is a specialist in nuclear medicine. Since 1994, he has been medical director of the Department of Nuclear Medicine at Milan's Ospedale Maggiore. He has written numerous international publications in the nuclear medicine field especially involving endocrinology.

Medardo Chiapponi is Professor of Industrial Design at the University Institute for Architecture of Venice (Istituto Universitario di Architettura di Venezia, IUAV). He has published books and papers in Italy, Argentina, Germany, Greece, and the USA, the most significant being: *Ambiente: gestione e strategia* (Feltrinelli, 1974); *Cultura sociale del prodotto. Nuove frontiere per il disegno industriale* (Feltrinelli, 1999); and, as editor, *Dalla cura delle cose alla cura delle persone* (Silvana Editoriale, 1999).

Alberta Contarello is Professor of Social Psychology at the University of Padua, Italy. Her main interests are nonverbal aspects of interpersonal relations, the social construction of knowledge (also with regard to the new technologies), and relations between social psychology and literature. Her latest publications include: (with C. Volpato) "Towards a social psychology of extreme situations: Primo Levi's *If This Is a Man* and social identity theory," *European Journal of Social Psychology* (1999), and (with B. M. Mazzara) *Le dimensioni sociali dei processi psicologici* (Laterza, 2000).

Elda Danese teaches textile design at the Venice Institute of Art (Istituto d'Arte di Venezia). She has organized several exhibitions, including "Woven Jewels: Jewel Fabrics" in 1998. She collaborates with the Center for the Study of Textile and Costume at the Municipal Venetian Museums of Art and History and in 2000 she wrote the text of the catalogue for the exhibition "L'arte al telaio: L'arazzeria Scassa dal 1957 ad oggi" (Allemandi, 2000).

Annalisa Dominoni is an architect and designer with a PhD in Industrial Design. She is lecturer in Industrial Design at Milan Polytechnic's Faculty of Architecture-Design and Head of Research Programs at SpaceLab, the research laboratory in space design, also at Milan Polytechnic.

Kim Hyo Dong is a doctoral candidate in the Rutgers University School of Communication, Information, and Library Studies, USA.

Leopoldina Fortunati teaches sociology of communication and mass media techniques at the University of Udine, Italy. Her main field of research is gender and information and communication technologies (ICTs) from a sociological point of view. She is part of the European programs COST269 (User Aspects of ICTs) and COSTA20 (The Impact of Internet on the Mass Media in Europe), and she is participating in the European research project SIGIS (Strategies of Inclusion: Gender and the Information Society). She has written several essays, including *The arcane of reproduction* (Autonomedia, 1995), and has edited *Gli Italiani al telefono* (Angeli, 1995) and *Telecomunicando in Europa* (Angeli, 1998).

Jorge Frascara is Professor and Coordinator of Visual Communication Design at the University of Alberta, Canada. He was president of Icograda (1985–1987) and has done pioneering work on research in graphic symbols for the International Standards Organization. He is the coauthor of *User Centered Graphic Design* (Taylor & Francis, 1997) and editor of *Design and the Social Sciences* (Taylor & Francis, forthcoming), and has published four books in Spanish. His current professional practice concentrates on traffic safety communications.

Paolo Gerundini is Professor of Nuclear Medicine at Milan's Ospedale Maggiore and Director of the Department of Radiological Sciences. He is a member of several scientific societies and is on the editorial board of the *Journal of Nuclear Medicine*. He also teaches in area schools, especially on the topic of medical radiology. His numerous publications are primarily in the field of cardiology, with a subspecialization in the diffusion of cerebral tracers among those suffering from degenerative diseases.

Paolo Girardi is a psychiatrist and teaches clinical psychology at the University of Rome, Italy.

Susan B. Kaiser is Professor and holds the Chair of Textiles and Clothing at the University of California, Davis, USA. She also has an appointment in the Women and Gender Studies program and teaches in the Cultural Studies doctoral program at UC Davis. Her research interests involve issues of style, fashion, and subjectivity, with a specific focus on intersections between age, gender, ethnicity, and sexuality. She is the author of *The Social Psychology of Clothing: Symbolic Appearances in Context* (Fairchild, 1997) and articles in journals ranging from semiotics to sociology and textiles.

James E. Katz is Professor of Communication in the School of Communication, Information and Library Studies at Rutgers University, USA. His latest book (with Ronald E. Rice) is *Social Consequences of Internet Use* (MIT Press, 2002). As part of the intellectual project presented in this book, he has also edited a collection of papers published as *Machines That Become Us: The Social Context of Personal Communication Technology* (Transaction, forthcoming). With Mark Aakhus he coedited *Perpetual Contact: Mobile Communication, Private Talk, Public Performance* (Cambridge University Press, 2002), which is the first multinational analysis of mobile phones. His earlier book on the telephone, *Connections: Social and Cultural Studies of the Telephone in American Life*, has won a prize from the American Library Association for being an outstanding academic title. Before joining Rutgers University he was director of social science research at Bell Communications Research. Katz has won numerous awards and grants, including postdoctoral fellowships at MIT and Harvard University. He holds two patents in telecommunications technology.

Richard Ling is a sociologist at the Telenor Research Institute in Norway. He received his PhD in sociology from the University of Colorado. He has become an internationally recognized expert in the field of new information communication technology and society. He has published numerous articles and has organized scholarly forums in the area.

Claire Lobet-Maris is a Professor in the Informatics Faculty of the University of Namur, Belgium, and is codirector of the Interfaculty Unit for Technology Assessment. She has conducted much important research into the use of ICTs. She was part of the European Research Project SLIM (Social Learning in Multimedia).

Giuseppe O. Longo is Professor of Information Theory in the Department of Electronic Engineering at the University of Trieste, Italy. He has carried out teaching and research activity in several universities and other scientific institutions in Europe, the United States, and Asia. Currently his scientific interests are mainly directed toward epistemology, artificial intelligence, communication, and the social impact of technology, topics on which he is very active in giving lectures and publishing papers. He is also a prominent writer of novels, stories, and other pieces.

Tomás Maldonado is Professor Emeritus at Milan Polytechnic, Italy, where in 2001 he was granted a doctorate *Honoris Causa* in Industrial Design. His wide-ranging interests as a scholar and teacher have revolved around philosophical and technical–sociological themes. His publications include *Il futuro della modernità*

(Feltrinelli, 1987); *Critica della ragione informatica* (Feltrinelli, 1997); and *Hacia una racionalidad ecologica* (Infinito, 1999).

Ada Neiger is Professor of the Sociology of Literature at the University of Trento, Italy, and researches contemporary Italian literature. She has recently edited the following books: *Il vampiro, don Giovanni e altri seduttori* (Dedalo, 1998); *Primo Levi. Il mestiere di raccontare, il dovere di ricordare* (Metauro, 1998); and, with W. De Nunzio and G. Pagliano, *Tracce d'infanzia nella letteratura italiana fra Otto e Novecento* (Liguori, 2000).

Virpi Oksman is working on the research project "Mobile Phone Culture of Children and Youth in Finland" at the Information Society Research Centre of the University of Tampere, Finland. She has a feminist approach and has written several essays on the use of mobile technology by Finnish adolescents and youth.

Giorgio Pacifici, a social and political scientist, is president of the Forum per la Tecnologia dell'Informazione (FTI), a cultural nonprofit association investigating the ICT scene, and president of the Associazione per la Ricerca Comparata e Interdisciplinare (ARCO), a research institute in comparative studies. As a professor he has lectured on Italian fashion problems at the Institut d'Études Politiques in Paris.

Anna Poli is Adjunct Professor in the Department of Industrial Design at Milan Polytechnic, teaching advanced-level courses on techniques of digital representation and computer-based design. She has been interested in visual design and advanced technologies since 1992. Her publications include (with Gui Bonsiepe) "Design, Wissen, Digitalisierung," in *Best Patterns Management* (Luchterhand, 2001) and "Designed Entirely by Computer," *Modo* (January 2001).

Pirjo Rautiainen has been working since January 1998 on the research project "Mobile Phone Culture of Children and Youth in Finland" at the Information Society Research Centre of the University of Tampere, Finland. She has written several articles on the mobile phone and Finnish children and teens. Her ethnographic research has also included American Indians, which took her to tribal homelands in Wisconsin, USA. Her research has been published in several edited collections, including *Machines That Become Us: The Social context of Personal Communication Technology* (Transaction, forthcoming).

Raimonda Riccini is a researcher in the Faculty of Design and Arts at the University Institute for Architecture of Venice (Istituto Universitario di Architettura di Venezia, IUAV), where she teaches history of science and technology. Her doctorate is in industrial design. She is a member of the Centro Studi Triennale di Milano and is coordinator of the Sezione Moda. A focus of her research is on the theory and history of technical objects, particularly household technology. Her articles and chapters on innovation, socialization, and propagation of technologies include "History From Things: Notes on the History of Industrial Design," *Design*

Issues (1998); "Innovation as a Field of Historical Knowledge," *Design Issues* (2001); "Il disegno industriale in Lombardia: Un modello per il 'Made in Italy,'" in *Storia d'Italia. Lombardia* (Einaudi, 2001); "The Appliance-Shaped Home. From the Electrified Home to the Wired Home," in *Italy. Contemporary Domestic Landscape 1945–2000* (Skira, 2001). She is the editor of *Il contributo del disegno industriale. Note di cultura politecnica* (Scheiwiller, 1999).

Dina Riccò has a PhD in industrial design (1996) and is Professor of Perception and Visual Communication at Milan Polytechnic, Italy. She was nominated for the 19th Compasso d'Oro (2001) for the book *Sinestesie per il design. Le interazioni sensoriali nell'epoca dei multimedia* (Etas, 1999). She is responsible for the website www.sinestesie.it.

Marco Somalvico, an electronics engineer, holds the Chair of Artificial Intelligence at Milan Polytechnic, Italy. He has dedicated his scientific activity to automatic problem resolution methodologies, visual systems, intelligent robots, intelligent multiagent systems, and virtual museums. He is the author of numerous international scientific publications.

Raimondo Strassoldo is Professor of the Sociology of Art at the University of Udine, Italy. His interests focus on the sociology of cultural and communications phenomena, and especially the sociology of the figurative arts and the impact of the new technologies on the production and consumption of art. He has published *Forma e funzione. Introduzione alla sociologia dell'arte* (Udine, 1998) and, as editor, the two-volume *Muse demotiche. Ricerche di sociologia dell'arte* (Forum, 2001).

Giovanni Strocchi is an electronics engineer and the Universal Mobile Telecommunications System (UMTS) Director of Omnitel-Vodafone.

Martha Turner is a doctoral candidate in the Rutgers University School of Communication, Information, and Library Studies, USA.

CHAPTER ONE

Introduction

Leopoldina Fortunati, James E. Katz, and Raimonda Riccini

1. THE BODY AND TECHNOLOGY AS CENTRAL SITES OF MODERNITY

The increasing integration of the human body with technology is a topic that has been attracting significant attention of late (for comprehensive reviews, see the works of Fortunati, 1995, 1998, and Maldonado, 1992, 1997, 1998; see also Brooks, 2002; Hayles, 1999; Katz, 2003; Kurzweil, 2000; Moravec, 1999). Scholars, social critics, lay people, and the press have weighed in on it. The topic also receives repeated attention from the world's religious, business, and political leaders. Indeed, it is likely to be one of the most significant social issues of the 21st century (Kelly, 1995).

Clearly the amazing progress in the integration of the human body with technological devices increases the anxiety of many, and the delight of others (Katz, 2003). But no matter what one's attitude toward the desirability of these developments, virtually no one regards them without passion. Whether the subject is genetic manipulation, plastic surgery, portable communication technology, intelligent household appliances, medical manipulation of the body, or more radical applications in the technologies of war, terror, and torture, all these issues require attention on both a practical and a political level.

It follows that this subject should also be an issue of scholarly interest, which it is, and an impressive output on the topic has accumulated. However, much of this scholarly work neglects the direct effects on the human body of technology's everyday advances. In light of this neglect, we have brought together a group of thoughtful scholars and analysts whose works comprise this collection.

The chapters in this volume discuss a wide array of topics impinging on how the body and technology interact. The perspectives include literary analysis, historical comparisons, analytical reports, and speculative interpretations. The purpose of all these efforts is to improve our understanding of the experience of body as it is mediated among competing forces and intellectual domains. Moreover, throughout the chapters runs a strand that takes up the question of what future lies in store for the body that has served humankind so well for so many millennia. As technology progresses, some fear that the body will become at best a mere appendix to the machine, at worst the machine's obliterated victim.

This fear is not out of place, because the body is increasingly running the risk of becoming an appendix to the machine, despite there being, as Nietzsche wrote (1973/1883–1885, pp. 34–35), more reason in the body than in our best wisdom. Technology, goes this line of argument, instead of being an instrument in the hands of human beings, is in danger of becoming (if it has not already become) its end, producing the inversion (and confusion) of the relative virtues of quantity and quality that, as Galimberti underlined (1999), Hegel had already described in his *Logic*.

As the illuminating analyses of Marx and then Simmel remind us, the first cause of these transformations of quantity into quality is money, which has become the dominant consideration in directing technological development. If in fact money serves as the force through which goods are produced and hence needs satisfied, it changes from being a means into an end. To attain it, we do not hesitate to sacrifice, when necessary, the production of goods (even going so far as to destroy products rather than see them sold or given away) and therefore reduce the satisfaction of needs.

Technology has suffered the same fate as money, even if it remains subordinated to it. From being a means, an instrument, it has become an end, to which even ethics and politics must bow in reverence. Ethics, which used to be where ends were described and elaborated, has become conditioned by technology, which obliges it to take positions on a reality that technology itself produces (bio-ethics) (Galimberti, 2000). Politics, from being a place where technology was assigned its specific ends, now adapts passively to technology. The result is that politics is less and less able to control or shape technology and therefore increasingly frequently acts only as a guarantee for it. At this point, politics runs the risk of assisting in the development of a society for technology and not with technology.

The body, however, as Heidegger (1959) suggested, seems to be fleeing from itself, because activity—the production of objects by the body—is a way for the body to possess itself in things, on the one hand, and to reveal the latent possibilities of things, on the other. For this reason, there is much of our bodies in technology, just as there is much technology in our bodies.

So the body is an entity with a high natural and cultural dynamics. And yet, in a world in which the artificial is extending its dominion over the natural, apparently without check, the body at the same time represents the maximum level of "naturalness" that is possible at a given historical moment. This double dynamics is even deeper than it seems if we look at the body in yet another way, that is, as a product, the result of a work process. The machine may be a "fille née sans mère," as Francis Picacia wrote, but human bodies have a mother (today even two).

Bodies in fact are produced by women in the domestic sphere. But, although bodies continue to be represented as natural products, they conceal an ever higher degree of artificiality. The sphere of reproduction indeed has not yet attained social visibility as the sphere of production of value, and there is not therefore a strong enough awareness of its central position as production. The domestic sphere, despite being where the continuous valorization of women's labor, and therefore of women themselves, is inscribed, is still a place of naturalness. Bodies as products of women in the domestic sphere seem therefore to be a natural product, which makes them socially quite defenseless and excludes them from clear and formal negotiation.

The human body, because it is placed outside the rationale of value, is seen as something whose value is so incommensurable and therefore immeasurable that it ultimately does not cost anything and so is socially devalued. As nonvalue, it therefore always costs less than technology (this is the reason for the slowing down of robotics).

Given these premises, the human body continues to be the place where the watershed between value and nonvalue exists as social differentiation, making the body increasingly impotent. And this is why technology is invading it more and more aggressively. The human body is undergoing the same processes today that nature once underwent. In fact, whereas initially technology turned to nature, today it has become very interested in the human body and is developing into technology of the body. There is one difference, however. Technology applied to nature served to reduce the expenditure of energy and human fatigue; its cost was thus absorbed by the industrial system, which was very keen to take advantage of technology's portentous applications. Technology applied to the body, on the other hand, does not start directly from an interest on the part of the system. It is individuals themselves who pay for turning to technology and they pay dearly. Processes that involve aesthetic or domestic technologies are almost exclusively individual or family processes, whereas those connected with health or communication also involve moments of collective negotiation, inside which the state becomes a strong decision maker.

This process of technologization of the body is generally depicted as a linear process, outside a social dynamics of conflict and political subjectivity. In reality it is a crucial political process that nevertheless finds it difficult to reveal itself for what it really is, for the precise reason that it involves those strategic places of social representation that misrepresent as natural what has not been that for a long time.

The body as natural technology has been falsely replaced today by technological innovation, which, as innovation, always presents itself as being better than what it replaces. If the same logic is followed with the relationship with the body of "right or inevitable progress," which is found in the development of technology, the body will be seen as something to eliminate. We must say with some bitterness that the limit, but also the fortune, of the body, at least for the time being, is that it does not cost anything and so is competitive with the cost of technological products (see robotics).

In the light of what has been said so far, a different political strategy seems to be becoming necessary around these processes. In this perspective, a fundamental role can be played by women as *bearers of another technological vision*. If we look at the Greek myth of Prometheus (in the version by Aeschylus), the gift of fire to humans ("pyrotechnics" or, literally, "fire know-how") was received not only by metallurgical man but also by woman, who, for example, transformed food from raw to cooked (Lévi-Strauss, 1964). If, in our semiparodic interpretation of the past, fire was used by metallurgically inclined "man" to forge arms for hunting and war (in the rationale of violence and domination), it was at the same time used by ceramically inclined "woman" as the technology of warmth, nutrition, and care, as recounted in the cult of the goddess Hestia (Bolen, 1984). (Significantly, *ceramic* is derived from the ancient Greek *keramos*, which means heat and earthenware.)

Therefore the path to follow is not only technological, but also and above all political. If, as Bacon wrote, knowledge is power ("Nam et ipsa scientia potestas est"), this power should be addressed in a new way, in the awareness that—to refer to the myth of Prometheus again—technology is accompanied by "blind promises": The great power of technology is not always accompanied by equally great visions. This is precisely why technology must be governed. We cannot go back, that is certain, but we must keep on moving in a different direction. This vague invocation is in its own way semiparodic too. But the urgency of addressing the issue becomes a clarion call when one considers the growing potential of hydraheaded biochemical and nuclear terror and war, on the one hand, and well-intentioned bodily invasions and subordinations from health and personal safety to performance enhancement and personal convenience, on the other. Indeed, the sophisticated insights and understandings offered by our authors suggest that humankind will be up to the task of dealing with these problems, that is, the metaphorical war, even as it also loses some important battles along the way.

2. THE HUMAN BODY BETWEEN TECHNOLOGY, COMMUNICATION, AND ART

In most places and at most times, people have enjoyed finding ways to blur the human body's physical boundaries. Clothing is the most prominent example, but jewelry, tattoos, and body modifications such as lip and neck extenders and foot compressors have all been used in disparate places and times. This blurring of boundaries seems even more prevalent today, owing to the complex merging of the artificial and the natural, and the confusion between "drug-free" states of being and states influenced pharmacologically (tranquilizers, mood elevators, and by-products of other medications, as well as regular stimulants and intoxicants); presence and absence are brought about by small personal technologies (Katz & Aakhus, 2002) and a profound alienation between mind and body. And of course there are the enormously popular techniques of plastic surgery.

Communication technologies have extended the boundaries of the body, increasing the capacity to transmit information. Technology has progressively grown closer to our bodies, approaching through first clothing, then synthetic clothing fibers, and finally "smart fabrics," wearable computers, and communicative machines embedded into jewelry, clothing, and even the body. The more prosaic technology of the 21st century includes mobile phones and laptops, which many over the age of 30 remember as being breakthrough concepts and machines in the 1980s but which are now relegated to the humdrum wallpaper of everyday life. Initially these technologies only approached the body; today's technology has fully invaded the body. This transition has set up a conflict not only with society's social and ethical perception of the body itself, but also with the aesthetics of clothing (see chap. 7 by Fortunati in this volume).

Fashion is an extension of both the physical and the aesthetic body—and is one of the first ways that technology began to enter the physical body's "space." Although today technology is usually thought of in terms of computers and cell phones, clothes and their material are also becoming "intelligent" (see chap. 16 by Danese in this volume). Human characteristics, such as intelligence, are attributed to objects such as clothing in order to make the environment more similar to humans. The application of "intelligence" in fashion evokes the same questions as the application of intelligence in communicative machines: Can an intelligent object exist that does not need to convey the passing of time with an increased need for beauty?

Human ideals of beauty expressed through fashion inevitably influence technological design. Today, communication technologies that reside near the body also embody taste and fashion. Indeed, fashion becomes central as the technology moves from being portable to being wearable (Calefato, 1996; Fortunati, 1998, 1999). Personal instruments of communication must be not only modern but beautiful. Communication technology and fashion aesthetics have recently come together to address this belief—fashion designers and telecommunications operators are increasingly combining their overall visions of aesthetics. Various research laboratories are developing clothing that will integrate communication, therapeutic, and cosmetic technologies.

Before the fashion industry was formed, society had long been accustomed to technologies of bodily modification, beauty, and medicine that manipulate and invade the body. Historically, nearly every society altered the body using the rationale of religion then medicine. In the West, modifications on religious grounds tended to diminish even as the medical rationales expanded. Lately, technologies that had already been used for diagnostic and therapeutic purposes are being expanded in order to explore and interfere with the body, invading it from both inside and out. Experimentation has already begun on ideas such as subcutaneous microchips to control domestic appliances and other electronically activated machines. The next steps are surely for expanded control over machinery, physical access to secure areas, personal identification, and location monitoring. "Quality-of-life engineering" creeps into everyday life through "intelligent house" technologies and artistic applications, as illustrated by Stelarc, the first artistic cyborg to use medical instruments, prostheses, robotics, virtual-reality systems, and the Internet to explore and increase the parameters of the body (Drey, 1997).

These technological "advances" result in an often welcomed assault on the human body, which is penetrated and manipulated for a wide variety of reasons. Increasingly, cosmetic surgery aims not simply to repair damage, to reverse signs of aging, or even to correct perceived aesthetic flaws in one's appearance. Rather, an element of creativity and fun has been introduced to allow for expressive modifications, such as strange or "unnatural" features.

As Longo has noted (1998, 2001), technology has recently developed even more quickly than science. The use of technological instruments is no longer subject to adequate testing and training, which were supposed to protect against mistakes in treatment. Putting technology into practice "without reflection," Longo stated, must be seen within the framework of technology's increasing distancing from argumentative thought caused by the complex (and therefore costly) field of scientific verification of the effects of technological innovation. It is no surprise that Gershenfeld (1999) openly admitted that the enormous social and industrial implications in research emerge more from the comprehension of what has already been implemented than from attempts to conceptualize it beforehand. Therefore, the effects of technology are discovered only years after their implementation. We see this, for instance, in the potential and real health and safety consequences of mobile phones and the social impact on the quality of social relationships (Katz & Aakhus, 2002).

3. PERSPECTIVES ON THE BODY

One element of the body that must be safeguarded is respect, which is closely connected to the identity of the individual. Ensuring that the body lives in harmony with the mind is part of this respect. Reactionary thought is always behind contempt for the body or its devaluation, as Maldonado (chap. 2 in this volume) quite rightly states. If one generation leaves behind a humiliated and profaned body, that generation becomes an accomplice to destructive processes.

The knowledge that has grown around these problems of technology and the body is currently quite advanced, but research and study often proceed in fits and starts. Segmented disciplines and elitist and noncommunicating rivulets develop, when the complexity of developments in this field require, if not exactly unified visions, at least convergence, confrontation, and discussion.

The purpose of the international conference held at the Triennale of Milan on January 11 and 12, 2001, was to try to understand the meaning of these processes. In particular, the aim was to understand the trends in communication technologies, fashion, design, and art and their impact on the body; what people gain and/or lose from this technological invasion of the body; what the possible risks are (anomie, loss of identity, strategies of social and individual control); and, finally, what the body itself is communicating.

In this spirit, scholars from different areas and disciplines—from artificial intelligence to the history of art, from semiotics to bio-technology, from anthropology to mathematics, from sociology to industrial design, and so on, many (seemingly) distant from one another—were invited to join in the debate. This meeting and confrontation of so many different opinions has taken on a greater urgency in the wake of world events. After the conference, and while the chapters were being prepared, the terrible assault on the United States was carried out by Islamicist terrorists. As this example shows, bodies become not only victims of war and other kinds of violence, but also weapons of war. Related examples can be seen in the Middle East, where suicide bombers use their bodies as the weapon casing. All these horrific events related to the use of the body cannot help but bring about even more profound thinking, research, and study along the lines reflected in this collection (and obviously many other lines as well).

This volume, which collects and makes permanent the work of the Milan conference, certainly reflects the variety of approaches. Our goal in this regard is to give many perspectives on issues, and, by inviting many juxtapositions, new facets of problems and issues should emerge. Our further hope is that this volume will be not a stopping point but rather a stepping-stone. It will help lay a foundation for a new perspective. Hence, we will be encouraging a breakdown of disciplinary barriers that have in the past prevented more effective use of the insights and creativity of scholars and analysts. After all, despite the fact of their physical embodiment, neither the body nor technology has a conceptual limit to what it can do or how it can be used (either in a positive direction or, sadly and horrifically, in the negative extreme). From this multidisciplinary, multinational base, we hope to help form a better base of knowledge, which is necessary to illumine the sense and dimension of this interpenetration between body and technology.

The main goal is for the many-voiced exposition of research results in different fields (epistemology, aesthetics, sociology, medicine, industrial design, and fashion) to trigger various kinds of connections and cross-fertilization. To foster this goal, we give a brief summary of the papers here, with an invitation to the reader to consider them in detail in the pages that follow. Then, in our concluding chapter, we seek to pull the variegated strands together to reflect on their larger meaning.

The first part of the book (in particular the chapters by Maldonado and Longo) concentrates attention on the body, the meeting point of processes of cooperation, competition, and tension between science and technology, and makes a notable epistemological contribution. Maldonado (chap. 2) reflects on the right attitude toward the body. The problem is finding the correct balance between making use of the prosthetic body, which is in any case the result of an act of faith in the human body, and abandoning the idea that the elimination of the human body is an inevitable stage in the development of science and technology. Longo (chap. 3), on the other hand, stresses the continuum between the body and technology. But to speak of *homo technologicus* as a hybrid between human being and machine also means addressing the maladjustments that this continuum involves, including the suffering that human beings are going through in today's bio-technological evolution, and also addressing Longo's general discomfort on the expressive, emotional, aesthetic, and ethical level. Fortunati (chap. 7) reconstructs the paths along which communication technologies have been able not only to approach the human body but even to penetrate it, without the body offering any resistance or refusal.

Somalvico's comments in chapter 4 are interesting in that he explores the relationship among humans, artificial intelligence, and robots from the perspective of both information engineering and anthropology. In particular, he explores how robotic architecture can be harnessed in the service both of people with standard capabilities as well as of those who are handicapped.

The often expressed social and psychological process of the body suffering pain (or, more properly, having pain inflicted on it by others) in the modern era of the graphical arts is discussed by Strassoldo (chap. 5). The human body, Strassoldo says, has been by far the most popular subject of the visual arts in the Western tradition. Body art has becomes the meeting place, or rather the place of fusion, between artist, technology, and artistic product. Neiger (chap. 6) provides a survey of representations of the body in literature, analyzing *topoi* (traditional themes in literary composition) of the description of the body of various characters in contemporary writers, such as Alda Merini, Max Scheler, and Malaparte.

The encroachment of technology on the body, especially communication technologies, is the factor that, according to many, indicates a new path of humanity. If previously technology's relationship was with places, today its relationship is with the body, as the site of the individual. The main technology that invades body has been identified as the mobile phone, the communicative instrument that has gone furthest down the path of individual encroachment.

Various aspects of the invasion and use of the mobile phone have been analyzed: how the cultural factor affects the perception of information and communication technologies (Katz, Aakhus, Hyo Dong Kim, & Turner, chap. 8); how the spread of the mobile telephone among adolescents has been dependent on the perception of what is "in" (Ling, chap. 10); how this spread serves to build social identities among groups of young people (Lobet-Maris, chap. 9); how children and adolescents see the mobile as a real extension of their hand (Oksman & Rautiainen, chap. 11). Contarello (chap. 13) addresses the importance of the simultaneous presence of bodies in communicative exchange and emphasizes the risks in terms of psycho-physical well-being when the body surrenders ever larger quantities of its own communicative ability to technological communication. These chapters illustrate the array of issues that are surely going to be exacerbated as more sophisticated communication technology becomes available. To help readers understand what capabilities are likely to be emerging, Strocchi paints a colorful picture of what the third generation of mobile communication is likely to entail (Strocchi, chap. 14).

Technology has come so close to the body that it has become technology to "wear" (Calefato, chap. 18), especially wearable computers (Poli, chap. 19). If technology is becoming wearable, this is also because fabrics themselves have become technological; that is, they enclose electronic parts and chemical substances (Danese, chap. 16). Communication and fashion are obviously the areas that come to the forefront, both on a profound level (Pacifici & Girardi, chap. 15), as well as from the perspective of social perception of anxieties that pass, rebound, and are manipulated by the media (Kaiser, chap. 17).

Comfort, well-being, and health are terms that have taken on increasingly rich and sophisticated meanings. Comfort today can no longer be seen only as the result of efficient services and environments; well-being and health no longer correspond only to the absence of problems or freedom from illness. Rather, all these concepts include a vast range of attributes in which the material and cultural, the physiological and anthropometric, and the medical and psychological elements are strongly interwoven. The human body is obviously the protagonist and recipient of this network of performances, with an increasingly pervasive role being played by technology. As has been observed several times, past generations have developed the belief that health, well-being, and quality of life are primary rights of human beings and as such they must be socially pursued and guaranteed. It is also increasingly the attitude that elevated standards of comfort, health, and well-being should be guaranteed by a higher standard of efficiency of technological service—and this is independent of the environment and the technical context in which services are distributed or requested (hospital, home, car, virtual reality, and even spaceships).

Although accepting these assumptions in principle, it is not possible to avoid certain basic questions that arise from the complex nature of the established relationship not only between technology and the body, but even more so between the technological and social systems. What is the space that public and private ethics can carve out for itself in the case of genetic manipulation? What are the risks of loss of identity in the direct conflict with ever more refined and powerful technology? What is the boundary between a reasonable amount of automation in the home and the illusion that total liberation from domestic work will become possible in that way? These questions are not purely rhetorical, but are real issues in this volume.

In particular, the papers by Jorge Frascara (chap. 22) and Raimonda Riccini (chap. 12) place the emphasis on negative aspects of technology. Without ever taking an antitechnological stand, these two papers investigate technology's capacity to preform, induce, or, conversely, condition personal and social patterns of behavior. The home and the car, both in their own ways at the center of individual identity, are the places where technology reveals itself to be inferior. On one hand, the car—a metal capsule full of aggressive symbols—becomes a magical shell for the body, a sheltered island, and, because of this, often leads to trouble. On the other hand, the technology of the home reveals its inadequacy in failing to do away with domestic chores and the rigidity of roles.

Other papers highlight cognitive, prepositional, and even planning approaches to the body and technology, dealing with issues of a more pragmatic nature. What are reliable procedures of intervention, improvement, and innovation in the field of health and well-being? How reliable are they? What kind of planning can help overcome the doubts and risks ensuing from the unrelenting increase in technological developments?

As Gerundini and Castellani illustrate in chapter 24, the representation of the body through images has taken on a central role in medicine. This is not only because it permits highly sophisticated diagnostics, but also because it appears to reveal information about the physio-pathological characteristics of human disease, and it supplies information on prognosis, the efficacy of the therapy, and the risks of side effects.

The rapidly increasing knowledge of the human body is paralleled by equally reliable technologies of cure and mechanical aid. One only need refer to the capacity of digital machines that approach the communicative models of human beings, acquiring the capacity of perception, or sensory data (Riccò, chap. 20); or the contributions of technological innovation in the field of textiles, allowing individuals to live and work in confined spaces or in conditions of microgravity in comfort and safety (Dominoni, chap. 23).

Above all, the latest research emphasizes the role of concrete products specific to body/technology interaction that also fit standards of well-being, efficiency, and functionality and maintain respect for the cultural and psychological characteristics of the user. For this reason it is important to emphasize the role of planning in the maintenance and recovery of the well-being of the body. A planning discipline such as industrial design can play a role in the production of sanitary items, medical, scientific, and analytical equipment, and diagnostic, therapeutic, surgical, rehabilitative, and support machines (Chiapponi, chap. 21).

REFERENCES AND BIBLIOGRAPHY

Bolen, J. S. (1984). *Goddesses in every woman. A new psychology of women.* New York: Harper & Row. Italian translation, *Le dee dentro la donna. Una nuova psicologia femminile.* Rome: Astrolabio, 1991.

Brooks, R. A. (2002). *Flesh and machines: How robots will change us.* New York: Pantheon.

Calefato, P. (1996). *Mass moda* [Mass fashion]. Genoa: Costa & Nolan.

Drey, M. (1997). *Escape velocity: Cyberculture at the end of the century.* New York: Grove.

Fortunati, L. (1995). *The arcane of reproduction.* New York: Autonomedia.

Fortunati, L. (1998). *Revêtir des technologies* [Wearing technologies]. *Réseaux, 90,* July–August, 85–91.

Fortunati, L. (1999). Abbigliamento tecnologico [Technological clothes]. In *Enciclopedia "L'Universo del Corpo"* ["The universe of the body encyclopedia"] (Vol. 2, 11–12). Rome: Istituto dell'Enciclopedia Italiana.

Galimberti, U. (1999). *Psiche e techne. L'uomo nell'età della tecnica* [Psyche and technics. The human being in the age of technics]. Milan: Feltrinelli.

Galimberti, U. (2000). *Orme del sacro.* Milan: Feltrinelli.

Gershenfeld, N. (1999). *When things start to think*. Cambridge, MA: MIT Press. Italian translation, *Quando le cose iniziano a pensare*. Milan: Garzanti.

Hayles, N. K. (1999). *How we became posthuman: Virtual bodies in cybernetics, literature, and informatics*. Chicago: University of Chicago Press.

Heidegger, M. (1959). *Gelassenheit* [Abandon]. Italian translation, *L'abbandono*. Genoa: Il Melangolo, 1983.

Katz, J. E. (Ed.). (2003). *Machines that become us: The social context of personal communication technology*. New Brunswick, NJ: Transaction.

Katz, J. E., & Aakhus, M. (2002). *Perpetual contact: Mobile communication, private talk, public performance*. Cambridge, England: Cambridge University Press.

Kelly, K. (1995). *Out of control: The new biology of machines, social systems and the economic world*. New York: Perseus.

Kurzweil, R. (2000). *Age of spiritual machines: When computers exceed human intelligence*. New York: Penguin.

Lévi-Strauss, C. (1964). *Le Cru et le cuit* [Uncooked and cooked]. Paris: Plon. Italian translation, *Il crudo e il cotto*. Milan: Il Saggiatore, 1966.

Longo, G. O. (1998). *Il nuovo golem* [The new golem]. Bari/Rome: Laterza.

Longo, G. O. (2001). *Homo technologicus* [The technological human being]. Rome: Meltemi.

Maldonado, T. (1992). *Reale e virtuale* [Real and virtual]. Milan: Feltrinelli.

Maldonado, T. (1997). *Critica della ragione informatica* [Critique of IT reason]. Milan: Felurinelli.

Maldonado, T. (1998). Ancora la tecnica. Un tour d'horizon [Technics again. A scan of the horizon]. In M. Nacci (Ed.), *Oggetti d'uso quotidiano. Rivoluzione tecnologiche nella vita d'oggi* [The use of objects in daily life. Technological revolution in today's life] (pp. 197–227). Venice: Marsilio.

Moravec, H. P. (1999). *Robot: Evolution from mere machine to transcendent mind*. New York: Oxford University Press.

Nietzsche, F. (1973). *Also sprach Zarathustra. Ein Buch fur Alle und Keinen* [Thus spoke Zarathustra: A book for all and none]. Italian translation, *Così parlò Zarathustra. Un libro per tutti e per nessuno*. In *Opere* (Vol. VI, 1, pp. 34–35). Milan: Adelphi. (Original work published 1883–1885)

PART I

The Body Between Science, Technology, and Art

CHAPTER TWO

The Body: Artificialization and Transparency

Tomás Maldonado

In 1855 Walt Whitman wrote, "I Sing the Body Electric" (1973, p. 93). The great apologist of bodily sensuality was using the adjective *electric* in the sense of exciting or thrilling. Today we might say "electrifying." But Whitman not only enthusiastically championed the body's grandeur. He also eulogized the landscapes of the nascent technical-industrial era. Consider his celebration, in a famous poem, of the beauty of the locomotive, describing it as a "black cylindric body, golden brass, and silvery steel" (1973, p. 471). It was an "emblem of motion and power," which Whitman saw precisely as a *corporeal* symbol of modernity.

Nevertheless, the "body electric" invoked by Whitman is not just a body capable, as it were, of electrifying us, of fascinating us to the point of enthrallment to its charms, its manifest or hidden beauties. Whitman was more probably thinking about a truly electric body—not only electrifying, but also electrified, a body unlike what had been known previously as the human body. We can imagine that the poet was interpreting the belief then widespread that a new human body, an *electric body*, was about to be born. This new body was destined to take advantage of the world of electricity that was then just emerging.

Although the 19th-century predictions of a body electric, *sensu stricto*, have proven to be more or less accurate, it would be more appropriate to speak of the *body electronic*. This is not merely a question of terminology. Upon closer examination, the new body we are currently concerned with (and about) is actually an electronically equipped body, a body assisted by microelectronic devices. The pairing of the body with electronics is often seen as a turning point in the history of the human body. This body will have become so artificial that some, perhaps with excessive nonchalance (as we see later in this chapter), even go so far as to speak of it as a post-human body. Nevertheless it would be a mistake to believe that the new body can be dubbed artificial only because of its capacity to incorporate electronic artifacts in its structure. Others factors play an equally important

15

role today. Examples include genetic engineering, assisted reproduction, and the use of psychotropic drugs. These technologies certainly are of enormous consequence, partly because they accelerate the process of artificialization of the body.

But artificialization in itself is not new. It did not begin with the use of these technologies. Even in the most distant past the human body manifested an innate tendency to become artificial, to be the object of artificialization. One thing is clear: a body in a "state of nature" has never existed since tribes were first organized.

In the final analysis the body has always been artificialized, because in one way or another the body has always been subject to the influence of culture. Therefore there is nothing new about the assumption that artificialization has played, and continues to play, a decisive role in the strategies of design (or redesign) of our bodies, strategies destined on an individual and collective level to meet evolutionary and environmental challenges. What is new, however, is the very high degree of artificialization that the body is achieving in our time. In this chapter, I focus on a particular type of artificialization, involving technical devices known as prostheses. These are added to (or inserted inside) the body to assist or substitute for an organ whose functioning has been impaired.

In recent times the body has become assisted by prostheses of all kinds. Nevertheless, the *prosthetic body*, the body that functions as a technical (or technicalized) subject, is significant not only in terms of performance or therapy but also in cognitive terms. This is for the simple reason that equipping a body with a prosthesis usually implies a notable increase in our knowledge, both of the organs in question and of the techniques required to produce the prosthesis.

Though the process of artificialization through prostheses dates back to the emergence of humans, ideas about the structure and functioning of the body long remained vague and superficial. In fact, with hindsight we see that most of them were wrong. The cognitive importance of the artificialization of the body has become evident only in recent centuries as awareness grew of the potential functional complexity of prostheses.

Understanding began to change radically when, during the Renaissance, the process of artificialization began to affect areas in which a more precise knowledge of the body became indispensable. In other words, the body could no longer continue to be a "black box." Of course, the efforts to reveal its secrets, to make it less opaque, did not immediately achieve results. In fact they have required centuries of work, and the process is still in progress.

Until the Renaissance, the only means available (to physicians as well as to others) were the senses of hearing, touch, and smell. Sight was important, but considered less reliable than the other three. It was not until the arrival of the great anatomists (and dissectors) of the Renaissance—Leonardo da Vinci, Berengario da Carpi, Andrea Casalpino, Andrea Vesalio, and Girolamo Fabrici—that vision acquired a central role (Bynum & Porter, 1993; Kemp, 1993). Vision was identified with dissection; it challenged the body's presumed holiness, setting out to make visible what is invisible by meticulously investigating the construction and the functioning of the works—the *fabrica*—of the human body (Camporesi, 1985).

This was the beginning of the reign of the eye—that is, visually oriented diagnosis of, and research on, the body. The great breakthrough came with the revolutionary discovery of X-rays by Röntgen. Medical radiology was born, as its name indicates, from a convergence of the physics of radiation and medicine. Yet Röntgen was not a doctor but an experimental physicist. Medical radiology is emblematic of the rapidly growing interdisciplinary nature of research and thinking in this area (Belloni, 1990).

Since the early 1980s, computer graphics modeling and simulation have yielded new insights for medical radiology. These developments affect diagnostic, therapeutic, and even surgical applications. They lead to research findings that enrich our knowledge of a universe previously concealed by somatic opacity, a universe that had given up few of its secrets, and then only through the most intrusive acts.

What remained unresolved was the problem of how to translate this knowledge into three-dimensional simulations or models to permit operative, interactive, real-time intervention with the bodies in question. Yet even this is being made possible by the new techniques of computerized medical radiology, as well as new virtual-reality systems that are beginning to bring the technical and biophysical systems together as an integrated whole. Thus medical imaging is enriched with new visualization tools and new solid modeling techniques. Suddenly we can *see* the organs and the apparatus of our body in four dimensions (three spatial and one temporal) and in multiple structures (mechanical, electrical, and biochemical). Moreover, it is also possible to intervene (even surgically) in the body's structures and functions.

The latest, and perhaps the most innovative, step toward absolute transparency of the body is the "Visible Human Project" developed in the United States. This project involves cross-sectioning frozen human cadavers in 1-millimeter slices. The images are digitally processed using tomography and magnetic resonance. And the results are astonishing: a three-dimensional map of the body of formidable realism. Using this interactive model of representation, previously impossible investigations of the human body are now feasible. Apart from its macabre aspects and ethical conundrums, the "Visible Human Project" has obvious advantages for diagnosis, therapy, and surgery.

Up to this point I have analyzed the development of the prosthetic artificialization of the body and examined the influence in this process of progress in the field of imaging techniques. Though we may express doubts and perplexities about many aspects of this type of artificialization—just consider, for example, the question of implanting prostheses at the cortical level—these steps forward are the result of an act of faith in the human body, an act of faith that, in spite of all its weaknesses and shortcomings, the body is capable, with the help of sophisticated technical auxiliaries, of improving its conditions of life and its relationship with the surrounding world. This act of faith is present even when the efforts fail or have undesired side effects.

The underlying choice has to do with the will not to renounce the care of the human body in the present and future. A human body is not an abstraction, but the concrete, everyday body that each of us occupies in a given historical moment. The human body has been *our* body for millennia, since the emergence of *Homo sapiens sapiens*. Certainly it is a temporary, vulnerable, imperfect body; a body never at peace with itself, constantly torn by conflicting experiences; a body that is a source of pleasure, but also of suffering; a body we love, and sometimes do not love; a body we wish were everlasting, but we know to be ephemeral.

Our body is not, as is commonly believed, what we *have*. It is—whether we like it or not—what we *are*. For this reason, and others, the defense of our body is, logically speaking, an obligatory choice. My perspective is that it is worth trying to ensure the continuation of the human species.

This is quite different from the position of those who see the possibility of creating an apocalyptic transcendental alternative to the human body in the progress of artificialization. This viewpoint is embraced by the exponents of "Body Art," science fiction, and artificial intelligence. They do not esteem the human body; in fact, the more indulgent view it with resigned, benign distrust.

Others disdain the body. For them, bodies are antiquated and obsolete. After millennia of stasis, suddenly the body should be replaced by another entity better suited for the pressing challenges imposed by an environment that is ever more strongly conditioned by new technologies.

Stelarc, an Australian artist famous for his imaginative bionic performances, wrote: "It is time to ask ourselves if a biped body, equipped with binocular vision and a brain of 1400 cubic centimeters, constitutes an adequate biological form." His response is negative, and he added: "There is no longer any advantage in remaining *human* or in evolving as a species, *evolution stops when technology invades the body*" (Stelarc, 1994, pp. 63–65; emphasis in original). Statements such as this are rather senseless, and in some ways even gloomy. But the media pay a lot of attention to them, and they seem to be gaining widespread credibility. In fact, many people, encouraged by the authoritative voice of Marvin Minsky, "think the body should be thrown away, that the wetware inside the skull, or namely the brain, should be replaced" (cited in De Kerckhove, 1994, p. 58).

I believe that what is at stake here, in philosophical and political terms, is too important for us to take such statements lightly. Such outrageous declarations of the need to throw the human body (brain included) onto the trash heap of obsolete objects raise the suspicion (or more than a suspicion, in my case) that they conceal Christianity's historical aversion to the body. But this time the aversion is garbed in a neo-mechanistic, sci-fi ideology.

There is little doubt that the prejudice against the body—the "abominable" body (Le Goff, 1969)—is a legacy that has profoundly influenced our relations with others and ourselves. Nietzsche (1960) had already glimpsed this, despising those who were "despisers of the body" (*"die Verächter des Leibes"*). This legacy must not be forgotten: Disdain for the body (especially that of others) too often leads to the brutal destruction of the bodies of women, men, children, and babies. The proof is there for all to see in the experience of the Nazi death camps, but also in the Inquisition, with its history of murder, massacre, and torture carried out by the "Holy Office."

So it is best to proceed with caution in this matter of a theory of an obsolete, inefficient human body worthy of being scrapped, as well as the idea of a body to be redesigned, based on some ideal model. This *biological essentialism* too takes us back to some very unpleasant memories. But although the theories of these modern "despisers of the body" may have, as we have seen, morally and politically abhorrent implications, this does not mean that the theme of the relationship between the body and technology is unimportant. To the contrary, it is crucial in hypermodern society. The problem is not the defense to the bitter end of the *natural* sacredness of the body, in the belief that moments of functional convergence between the body and the technical cannot exist (after all, they have always existed).

Certainly the borderline between natural life and artificial life appears to be increasingly elusive today. The thesis put forward 30 years ago by Georges Canguilhelm (1971) about the *continuity* between life and technique, between the organism and the machine, seems to be meeting with definitive confirmation. We do not have *androids* on the one hand and *nonandroids* on the other. Exchanges between them are now intense and frequent, and phenomena of (near) hybridization and symbiosis are the order of the day.

After all, the body has always been conditioned (and even determined and formed) by sociocultural techniques. Just consider Marcel Mauss' "techniques of the body" (Mauss, 1968) and Michel Foucault's social (or practical) techniques of coercion applied to a body that has become an object, a "body-object" (Foucault, 1975). The former explain how people, in any society, know how to make use of their own body; the latter explain how people, in any society, make use of others' bodies for their own ends.

We have already discussed how the artificialization of the human body is conditioned today by the growth of new technologies. Within this theme, I would like to probe certain questions that have emerged in the field of robotics.

Many experts in this field, although they do not advocate total substitution of the human body (an approach supported, as we have seen, by the champions of the "cyberbody"), are convinced that some of our sensory faculties could be made more efficient through the use of artificial sensors. In particular, they are interested in the artificial tactile sensors that are fundamental parts of robots designed for manipulation and assembly functions.

Recently great progress has been made in the artificialization of touch (tactile sensors and effectors). In the overwhelming majority of cases these are devices that involve pressure and the "retroactivity of force." But when attempts are made to develop synthetic skin and fingers capable of *feeling*, for example, the smooth or rough surface of an object, the results are not even remotely comparable to the performance of natural skin and fingers. One very important point is usually overlooked: A person's natural sense of touch does not consist only of contact; touching is not just touching. Our sense of touch perceives multiple factors even without true direct contact with our skin.

That skin that covers the entire surface of our body is not a passive wrapper that protects us from the external environment and separates us from the world. The skin is one of the most effective mechanisms for interaction with the world. It is the locus of a very wide range of types of sensitivity. "The skin has eyes," Diane Ackerman (1990, p. 94) wrote, using an exaggerated but apt metaphor.

To stretch the metaphor even further, we might add that the eyes have skin. In other words, sight can be understood as an extension of touch, but touch can also be an extension of sight. We have to admit that today's artificial touch is but a pale imitation of natural touch. And, given the present state of our knowledge, it is hard to imagine how the situation could change in the future.

In conclusion, I would like to return to Stelarc's declaration that *"evolution stops when technology invades the body."* Here we can clearly glimpse the thesis, also supported by Moravec (1988), that the human species is destined to disappear and in its place, sooner or later, a new species will appear, a posthuman or nonhuman species. The advocates of this position have not indicated exactly what bodily form the new posthuman species might take. They usually prefer to remain rather vague regarding such matters. However, the most uninhibited of them suggest that the appearance of the posthuman body will be very similar to what we have recently become accustomed to in science fiction novels and horror movies, namely the image of "cyborgs" or androids.

One among the many arguments used to support this disconcerting scenario deserves particular attention. I refer to the assertion that the defense of our species is prompted exclusively by the continuing existence of an affected anthropocentrism, a stubborn refusal to admit that our species, like any other species, might vanish to facilitate the appearance of a more advanced one. This appeal to anthropocentrism, in this context, has implications that should not be underestimated.

We all know that the end of the geocentric cosmogony, thanks to Copernicus, Kepler, Galileo, and Newton, also marked the end of anthropocentrism. Their work inadvertently led to the downfall of the idea, deeply rooted in the theological/religious tradition, that human beings, as inhabitants of an earth believed to be the center of the universe, could also be considered to be at the center of that center. With the collapse of this worldview, a relativism emerged regarding the role of our species in the universe. This relativism has had particularly fertile results in a wide range of fields of knowledge, including anthropology. Since World War II, in fact, relativism has been the dominant orientation of anthropology, especially of cultural anthropology.

Nevertheless, it must be noted that certain illustrious exponents of cultural relativism have recently shown signs of impatience with its overly dogmatic, reductive version. For example, Clifford Geertz (2000) hypothesized a cultural relativism with new basic tenets, a relativism that uses some of its adversaries' pertinent critical objections to develop a more subtle, articulated version. He called this "anti-anti-relativism." In my opinion, such critical revision should also be applied to antianthropocentrism, and in particular to the caricature-like forms it has assumed in our time in the wild, obscure theories that have been concocted regarding a future posthuman species.

I want to be clear: I am not suggesting that, out of principle, we should exclude prehuman and posthuman worlds from scientific and philosophical study, worlds in which we humans, as a species, have been or will be absent. Nothing could be more irrelevant to my way of thinking than the obtuse anthropocentrism that for centuries obstructed the development of knowledge in every fields. I nevertheless feel it is understandable, if not justifiable, that we humans, as members of the anachronistic human species, are not highly motivated to work on the advent of a posthuman species, a species that would, moreover, have the job of replacing us. I believe it is our more than legitimate right not to get involved, any more than is necessary, with promises of possible (or impossible) worlds in which the life of our species is excluded.

By raising this objection, I know I may be suspected—perhaps with some justification—of accommodating anthropocentrism. But I am convinced that a certain amount—a very small dose—of the old anthropocentrism might help restore our hopes regarding the fate of our body, and therefore of our *humana conditio*, at a time when the gloomiest prophecies about its future are tending to capture the attention of a growing audience.

All things considered, anthropocentrism contains a nucleus of good sense, which it could be useful to recover in the face of the excesses of antianthropocentrism. The Italian physicist Toraldo di Francia observed: "We might as well recognize the fact that it is biological—and therefore natural—law that mice are *mousecentric*, cats are *catcentric*, and men are *anthropocentric*. Otherwise ... no species would be saved from extinction" (1991, p. 21).

REFERENCES AND BIBLIOGRAPHY

Ackerman, D. (1990). *A natural history of the senses.* New York: Vintage.

Belloni, L. (1990). *Per la storia della medicina* [Towards a history of medicine]. Sala Bolognese, Italy: Arnaldo Forni Editore.

Bynum, W. F., & Porter, R. (Eds.). (1993). *Medicine and the five senses.* Cambridge, England: Cambridge University Press.

Camporesi, P. (1985). *Le officine dei sensi* [Workshops of the senses]. Milan: Garzanti.

Canguilhelm, G. (1971). *La Connaissance de la vie* [The knowledge of life]. Paris: Vrin.

De Kerckhove, D. (1994). Remapping sensoriale nella realtà virtuale e nelle alter tecnologie ciberattive [Sensory remapping in virtual reality and other electronic technologies].In P. L. Capucci (Ed.), *Il corpo tecnologico. L'influenza delle tecnologie sul corpo e sulle sue facoltà.* [The influence of technology on the body and its faculties]. Bologna, Italy: Baskerville.

di Francia, G. T. (1991). *Un universo troppo semplice* [A too simple universe. Milan: Feltrinelli.

Foucault, M. (1975). *Surveiller et punir. Naissance de la prison* [Discipline and punish: The birth of the prison. Paris: Gallimard.

Geertz, C. (2000). *Available light. Anthropological reflections on philosophical topics*. New York: Princeton University Press.

Kemp, M. (1993). The mark of truth: Looking and learning in some anatomical illustrations from the Renaissance and eighteenth century. In W. F. Bynum & R. Porter (Eds.), *Medicine and the five senses*. Cambridge, England: Cambridge University Press.

Le Goff, J. (1969). *L'imaginaire médiéval* [The imaginary medieval]. Paris: Gallimard.

Mauss, M. (1968). *Sociologie et anthropologie* [Sociology and anthropology]. Paris: Presses Universitaire de France.

Moravec, H. (1988). *Mind children. The future of robots and human intelligence*. Cambridge, MA: Harvard University Press.

Nietzsche, F. (1960). Also sprach Zarathustra [Thus spoke Zarathustra]. In *Werke* (Vol. 3). Munich: C. Hanser Verlag.

Stelarc (1994). Da strategie psicologiche a cyberstrategie: Prostesica, robotica ed esistenza remota [From psycho to cyber strageties: Robotics and remote existence]. In P. L. Cappucci (Ed.), *Il corpo tecnologico. L'influenza delle tecnologie sul corpo e sulle sue facoltà*. Bologna, Italy: Baskerville.

Whitman, W. (1973). "I Sing the Body Electric" and "To a Locomotive in Winter." In S. Bradley & H. W. Blodgett (Eds.), *Leaves of Grass*. New York: Norton.

CHAPTER THREE

Body and Technology:
Continuity or Discontinuity?

Giuseppe O. Longo

1. BODY AND TECHNOLOGY

When we consider the problem of the relationship between humans and technology, we often tacitly assume that humans and technology are distinct and separate entities. Moreover, we assume that technology evolves very rapidly, whereas human evolution is very slow or even static. My premises here are quite different: On the one hand, the distinction between humans and technology is not sharp, because technology has always had a big role in shaping the intimate nature of humans, and, on the other hand, technology's evolution has gradually taken the place of humans' evolution and has become a sort of continuation of it.

These two evolutions have become closely intertwined and have formed a "bio-cultural" or "bio-technological" evolution that has set the stage for the appearance of a new species, *homo technologicus*, a symbiotic creature in which biology and technology intimately interact.

The bio-technological evolution is ruled by a mixture of Darwinian and Lamarckian mechanisms and forms a composite tangle, which becomes even more complicated as the human–machine symbionts connect to each other to form a sort of global (cognitive) organism heralded by the Internet. As the Internet develops a sort of connective intelligence of its own, some scientists maintain that we are approaching a posthuman era in which intelligent machines will usurp the role of humans as stewards of the planet—and that with the cooperation of humans themselves!

Homo technologicus is not simply "*homo sapiens* plus technology," but rather "*homo sapiens* transformed by technology"; it is a new evolutionary unit, undergoing a new kind of evolution in a new environment. The novel symbiont is immersed in the natural world, hence obeys its laws, but also lives in an artificial environment, characterized by information, symbols, communication, and virtuality.

We know that the mechanism of inheritance of acquired characters proposed by Jean-Baptiste Lamarck to explain biological evolution does not work, because it would soon lead a species into an evolutionary dead end. But it does work in cultural evolution, where imitation is a powerful and quick mechanism that shortcuts the long and slow selection process proposed by Charles Darwin to explain biological evolution. Owing to Lamarckian mechanisms, technological (or cultural) evolution is rapid, because it lacks negative feedbacks to slow it down: Immediate adaptations to novelties tend to root immediately into the deep structure of society. The rapidity of this evolution, however, also causes it to be fragile.

In the 20th century, technology underwent important changes. First of all, the new "information technology" appeared alongside the old technologies dealing with matter and energy, and mind-machines were built (i.e. systems for processing, storing, and transmitting data). Second, after exploding into the world to modify it, technology, especially information technology, has begun to implode, invading the body and spreading throughout it with a variety of micrometric and nanometric devices able to modify cells and even molecules, with consequences that are difficult to assess.

True, this is to some extent a *scenario*, but today we tend to resort to this semifictional method to investigate the present situation and future developments, because the classical instruments of rational prediction are not very efficient in a highly complex world that is evolving ever more rapidly. It is as if our future were the subject of a *narrative* in which science and fiction unite to yield plausible predictions.

If the scenario I am sketching here is correct, and if *homo technologicus* does emerge from history and evolution, what are the consequences likely to be? I am interested in the discontinuities and in the mismatch between the organic and the artificial components of the symbiont, rather than in the continuity aspects of the shift from biological evolution to bio-technological evolution. No doubt, the two evolutions are somewhat heterogeneous and, consequently, the two components of the human–machine hybrid are heterogeneous. This mismatch could cause certain kinds of suffering that would add to those resulting from our organic nature, although technology has contributed to relieving some "conventional" forms of pain.

The deepest human characteristics, those associated with the emotions, communication, expression, the atavic inheritance linked to the body and rooted in the most ancient layers of evolution, which played a fundamental role in the survival and development of our species, would not immediately disappear just because technology had inserted its nanometric prosthetic devices in our bodies and brains. And in the area of contact, in the infinitesimal interface between "us" and "our" prostheses, serious rejection processes could appear. Even today, when *homo technologicus* is still at an early stage of development, we can observe problems and difficulties arising from the mismatch and incompatibility between human and machine. As evidence of this, note that a great deal of research effort is being devoted to the construction of user-friendly machines, which should create an anesthetized zone into which the artificial components are allowed to sneak. In other words, we purposefully try to weaken the resistance of the ancient body components to the encroachment of the newer mind constructions. The consequences of this course of action are difficult to predict, but it could be a source of problems.

In the technology cage that we are building around ourselves like a tight suit, some of our skills will be as useless as prehistoric relics, but will nevertheless continue to demand to be put to use or will ache like phantom limbs. Other skills will obviously be enhanced: Technology will operate a sort of selective *filtering* on our person (the complex unit of mind and body).

All these considerations concern in particular our bodies. The body is spotlighted by the informational revolution: Electronics, robotics, and spintronics invade and transform the body and, as a consequence of this, the body becomes an *object* and loses its remaining personal characteristics, those characteristics that might make us consider it as the sacred guardian of our identity. The body has varied in popularity throughout the ages, but overall it has been viewed rather negatively. Many Greek philosophers disparaged and despised it, considering mind and soul much nobler than the body; Christianity mistrusted the body because of its strong leaning toward sin; Descartes and all his followers considered it as a mere support for the nobler mind. Today, finally, the body is on display on the stalls of the global market, where its parts are bargained for and sold. And all these transformations occur under the strong thrust and impetus of the economy.

The current status of the body is rather confused and contradictory: On the one hand, it is now recognized that the body is the robust container of intelligence and of implicit and primary knowledge, as opposed to the fragile and abstract mind postulated by symbolic artificial intelligence; on the other hand, the body is still considered inferior and bodily reproduction is considered, as always and everywhere, cost-free and obvious. Doctors and biologists, engineers and technicians, use the body to perform transgressive and amoral experiments that for some will take us toward a wonderful demigod-like future and for others are simply a profanation and debasement of what is most intimate and individual to each of us. Preserving the body, defending its integrity, and seeking its well-being within a finite and harmonious temporal horizon are countered today by eugenic attempts to defeat diseases and even death through systematic recourse to technology.

At the same time, the body is the object of a morbid and almost pathological interest centered on an obsession with exercise, massage, plastic surgery, piercing, warping, twisting, crippling, maiming. The manic, often even abnormal, attention to the body, from makeup to depilation to cosmetic surgery, is a conspicuous aspect of the artificial and has spawned a vast business. All this goes with an irreparable devaluation of the "natural," or rather with a progressive *confusion* between the natural and the technological, so that the body is integrated with, and replaced by, mechanical parts. Machines are felt to be the real trustees of incorruptibility, of imperturbability, and perhaps, tomorrow, of immortality, and we eagerly want to become machines or machinelike. The body becomes the object of artistic experiments, becomes a spectacle and theater. Its anatomy, functions, and organs are disaggregated and observed analytically with a view to curing, correcting, and modifying them, enhancing or suppressing them, to free the body utterly from the legacy of bio-evolution.

The effects produced on the body by information technology are particularly interesting. *Telematics* and *virtual reality* produce a communicative, perceptive, and functional *diffusion* of the body. Virtual reality, for instance, extends the body spatially in unprecedented ways and allows it to occupy the whole planet. Distance is annihilated and sensitivity is dis-located, but, paradoxically, by denying the body's primary attribute—its proximity or presence—through artifice and simulation. Virtual reality enhances the body by denying it: You can travel a long way without leaving your room, without enacting the space–time dislocation that is (was) necessary for the body to perceive, and consequently to exist. Tele-actions imply a tele-existence and the space–temporal world of reality is lost and replaced by a *tele-space–temporal* world that can be manipulated at will. Virtual reality gives us substitution techniques that may lead to ubiquity but at the same time weaken and reduce (according to some, *kill*) immediate perception. Omnipresence and complete inertia go together. Moreover, virtual reality permits us to take on extravagant or chimerical identities, so as to offer an arbitrary personality to our partners and interlocutors.

Consequently, the "given" reality also takes on virtual characteristics: We are immersed in a world with which we interact, and such interactions modify both the structure of the world and ourselves. In this interactive dialectics, the subject and the world are basically one thing and constitute themselves simultaneously, so that we cannot talk of *one* reality or *one* world given once and for all; but instead we should talk of *many* realities or worlds that appear different to different eyes and that (as in quantum mechanics) constitute themselves in the very moment in which the interaction that evokes them takes place.

Again, this separation of the body from its primitive functions and this weakening or negation of presence induced by technology imply pain. This is why children, even more than adults, should be given the joyful possibility of full immersion, body and mind, in the intense and healthy experience of being alive and *in* the world. A hypertrophic development of the mind, as is induced by computers, rather than direct experience and physical interaction with other living beings, leads to a flattening of human capabilities, even of the most abstract ones, such as verbal skills. We live and develop through interaction with other people, and such interaction is the more efficient and nourishing the more it is linked to the rich and complex characteristics of bodily expression.

Giving up or repressing the body would lead to a serious impoverishment of our communication skills, which are very finely tuned and give us so much satisfaction. As fragments of a vast communication system, humans have an inborn bent for communication, sign interpretation, linguistic interplay, lies, theater, acting, recitation, and so on. Communication is not primarily a cognitive or conceptual experience; it is a global activity of the unity of mind and body that we call a person. We talk, tell stories, argue, and perform, and this continuous and diffuse communication activity is based on our original body–mind nature, which communicates even before we communicate explicitly.

Information technology introduces drastic simplifications and mediations that tend to abolish distinctions and nuances in this diverse and complex landscape. This facilitates human–machine communication, and therefore it is useful for exploiting instruments. At the same time, however, it impoverishes human communication: People tend to use a more restricted vocabulary and to resort to a limited number of grammatical and syntactic structures. And the body disappears. As the vehicle simplifies, so expression and communication risk becoming a bunch of rigid clichés.

2. CREATIVITY, ETHICS, AND AESTHETICS

To which of the two components of the human–machine symbiont should we attribute the various active and cognitive functions of *homo technologicus*? We could beg the question by saying that the two components are inseparable. Furthermore, the question could be considered a false problem, because it is the symbiont as a whole that possesses creativity. This problem is similar to the issue of whether or not machines can think: In a sense the question is irrelevant, because it is always the human–computer unity that thinks. True, intelligence always stems from the message exchange and an intense communication process takes place in the computer as well, but computer messages have no meaning until they are interpreted by the human component (as happens with the Internet, whose messages have meaning only because there are humans at the terminals). Actually, the actions of human beings have a *meaning* and are reflected by *consciousness*, the meaning being rooted in the world and history and coming before the actions, whereas consciousness yields the subjective dimensions of the self in an elusive yet indisputable way.

Going back to creativity, humans have always been inseparable from their technological instruments, and this is true also in the case of *homo technologicus*. Informational instruments "filter" creativity as they filter perceptions, cognitions, and actions. Like all filters, the computer filtering interface is selective, because it enhances certain capabilities and weakens or suppresses others. In other words, our being in the world, and hence meaning and consciousness, are filtered by instruments. As a consequence, technology modifies creativity, changes its characteristics, and introduces expectations and perspectives that would not otherwise be present. You have different expectations depending on whether you are going to paint with a brush or use a computer to create fine-art works.

Moreover, the computer can run programs to perform some parts of the art work automatically—the artist need not finish up all the details of the work, but can assign them to the computer, just as some great Renaissance artists used to entrust their disciples with the finishing of their masterpieces. Such a distribution of tasks could also take place in literature or poetry (think of the function of such conventional tools as thesauruses or rhyming dictionaries). In other words, one can conceive of an artistic externalization as much as of cognitive externalizations (computers, databases, dictionaries, and the like). In the future, such artistic externalizations could contribute to the diffusion of creativity, shifting it partly from humans to machines, once the latter become flexible enough. Much depends on how the human–machine interface develops and on the quality of the interaction between the two components.

In the artistic field, as in the scientific, technical, and mythological fields, humans tend to reconstruct a part of the natural world, filtering it through their perception, sensitivity, and cognition. The prime instrument of this reconstruction is the body, with all its perception, cognition, and action organs. As the body becomes enriched with instruments and prostheses, however, the reconstruction is performed through those added peripheral devices as well. Whenever *homo technologicus* acts in the artistic, scientific, or technological worlds, he also tries to reconstruct the natural world, resorting to all his perceptive, cognitive, and active instruments. Such instruments derive from an integration of the tools of the two components of *homo technologicus*, but primarily the biological body.

In the case of *homo sapiens*, the reconstruction of the world, be it artistic, scientific, or technological, obeys criteria of economy or even survival. But, in art and poetry for example, emotional, expressive, ethical, and aesthetic components also come into play, as well as symbolic and religious needs that cannot easily be reduced to material or instrumental considerations. Moreover, artists want to express their sadness or joy, the uneasiness of being, the vague and ill-defined anxieties that stem from contemplating the night sky or distant landscapes, the mystery of being alive, the terrible events of birth and death.

Those components and needs have their roots in biological evolution and stem from the immersion in the world of the unit of mind and body that we call a human being. To what extent can the new symbiont share them? There are in fact great differences between the artistic activity of *homo sapiens* and that of *homo technologicus*. Maybe it is too early to draw a picture, albeit approximate, of such differences, but some of them can be hinted at by the following questions: What is the *aesthetics* of the symbiont? What is its *ethics*? What are the links between the two? What world does the symbiont (re)create?

Such questions are of paramount importance for the low-technology human, for whom ethics and aesthetics are closely linked together, because both are rooted in evolutionary history and both mirror life and evolution. In a sense, ethics and aesthetics coincide (or are isomorphic) because they stem from the strong coimplication of species and environment: *Aesthetics is the subjective feeling of our harmonious immersion in the environment; ethics is the subjective feeling of respect for, and harmonious action in, the environment of which we are a part.* Consequently, ethics allows us to preserve aesthetics, and aesthetics serves as our guide in ethical action.

Ethics and aesthetics are historical; that is, they evolve, both at the species level and at the individual level: experiences in varying contexts produce ethical and aesthetic novelties that seem to have a concrete physiological counterpart in the activation of specific brain circuits. As a consequence of the Cartesian and Baconian separation between a human being's components and between humans and nature, and of scientific thought and technology, ethics (i.e., the complex of "just" courses of action, able to grant a dynamic, harmonious survival) has been subject to enormous stress for a long time, and this stress is today at its strongest. It would appear this has had important consequences for aesthetics as well.

Ethics and aesthetics are modified also by the strong *simplifying* effect that technology has on our image of the world and of humans. All this has brought about a serious crisis in aesthetics, to which the process of abstraction and coding that underlies scientific formalism has contributed. In contrast to natural messages, humans' signs and codes are *arbitrary*. In music, in architecture, in painting, and partly also in literature, aesthetics has been disrupted: the harmony between humans and their evolutionary immersion in nature has been replaced by the *combinatorial arbitrariness of signs*, as is demonstrated by many of the recent trends and modes in music or painting.

The cognitive externalization represented by computers and by the Internet has changed the terms of the ethical–aesthetical problem. What "brain" connections in a general sense (biological or artificial) are activated by experiences? How do these new connections interact with the totality of preexisting connections? In the case of the symbiont, can we say that "life" is mirrored by aesthetics and ethics? Can we speak of "harmony" or "coincidence" between ethics and aesthetics or of "respect" for the vital or biological needs of the global system? How does the symbiont "evolve"? And in what "environment" or "world" does this evolution take place?

These questions may sound too "philosophical," but they are important because they go to the very roots of our relationship with machines and of our bio-technological evolution and they shed some light upon the fundamental importance of that elusive, mysterious entity that we call the body.

REFERENCES AND BIBLIOGRAPHY

Caronia, A. (1996). *Il corpo virtuale* [The virtual body]. Padua, Italy: Muzzio.
Longo, G. O. (1998). *Il nuovo golem* [The new golem]. Bari/Rome: Laterza.
Longo, G. O. (2001). *Homo technologicus* [Man, the technological species]. Rome: Meltemi.

CHAPTER FOUR

Bodies and Robots

Marco Somalvico

The aim of this chapter is to show the connections between our bodies and robots, and more generally to illustrate these relationships both in the sphere of information engineering—in the light of robotics and artificial intelligence—and in the sphere of philosophy—in the light of anthropology. I place special emphasis on the notion of robot architecture, with detailed reference to its use to serve both the disabled and by extension the able-bodied.

Recent developments in robotics also are illustrated, such as those in the sphere of anthropic robotics. Anthropic robotics refers to two aspects of the application of robots: (a) robots worn directly on the human body and connected to the outside; (b) in the future, robots inserted inside the human body to carry out functions of the disabled subject, and to strengthen and at times integrate or automate functions in able-bodied subjects too.

I therefore give some indication of the integration that could develop in these new fields, such as the by now classic field of domotics, which refers to the use of computers and robots dedicated to automation in the home (*domus*), and the more recent and still emerging fields of urbotics and orbotics, which refer to the use of computers and robots dedicated to the automation of the city (*urbs*) and the land (*orbis*), respectively. These new fields of interest are in part interconnected and interwoven with the new "ubiquitous computing," which can be seen in the progressive insertion into the body of computers and robots connected by electromagnetic waves and so not requiring wires. From this perspective, computers and robots (or, respectively, information agents and robotic agents), when connected to one another, constitute a new kind of machine, called an "agency" or "multiagent system." We can conceive of this as a "communication machine," which, depending on the different agents belonging to the agency, is called an anthropic, domotic, urbotic, or orbotic agency.

Finally we may observe how sometimes an artifact, to be precise a machine (in our case robotic agents and information agents), whether understood as an anthropic, domotic, urbotic, or orbotic agent, is able to supply services not available among the skills of a body of a disabled subject, or is able to strengthen and integrate these skills with those offered with greater efficacy by modern information, communication, and cooperation technologies. The prospects of developing new information and robotic tools, and their likely success in providing aid to the bodies of both disabled and able-bodied subjects, are therefore to be seen as a creative approach to coping with a broad range of situations and as a stimulus for increasing individual productivity. In substance, the new information and robotic tools are tools of a new kind, valid for the bodies of all people.

It would therefore seem useful to provide a critical synthesis of the potentialities offered by modern information, communication, and cooperation technologies. These are to be seen as of general service to all people, who are always in search of new modes of communication and socialization within modern human society.

In particular, I wish to emphasize a vision of an individual (termed "IT [information technology] and robotic bipolar"), understood as a subject that is human/mind, characterized by the copresence of a human/body pole and a human/machine pole. These two poles are seen as active and reactive, respectively, and as belonging to a single interpersonal (or, rather, intrabipolar) communicative and collaborative interior process. The artificial skills of the human/machine pole obviously compensate for the lack of human skills in the human/body pole of a disabled subject, who, understood globally as bipolar, is therefore no longer at a disadvantage (the true meaning of handicap). It is furthermore obvious that the skills supplied by the human/machine pole are not infrequently equivalent from a qualitative point of view to the skills supplied by the human/body pole of an able-bodied subject and also more efficacious from a quantitative point of view.

This has the really paradoxical result that able-bodied subjects who decide not to replace the skills available in their human/body pole with more efficacious ones available in their human/machine pole will find themselves, despite being able-bodied, handicapped in respect to disabled subjects.

1. UBIQUITOUS COMPUTING, ANTHROPIC AGENCIES, AND NANOTECHNOLOGIES

In the future development of information technology and robotics, with the gradual shift from microelectronics to nanoelectronics, agencies will take on an increasingly important role (remember again that an agency or multiagent system, considered as a cooperation machine, is an artificial system made up of several agents, each corresponding to a single computer or robot). At the international level this role is called "ubiquitous computing," that is, the concentrated agency, based on the local web or the Intranet (in the case of domotics, the automation of the building with a series of information and robotic agents), or the distributed agency, based on the geographical web or the Internet (as in the case of electronic commerce).

This new role for IT and robotics, distributed by means of the Internet and robots, is based on the progressive and unstoppable insertion of artificial devices—

not just aids, ortheses, and prostheses, but real information machines (both computers and robots). What can be envisaged therefore in the midterm is not only the insertion of the bipolar human being inside "ubiquitous computing" but even the insertion inside the body of the human being of a multiplicity of information and robotic agents, called an anthropic agency.

The anthropic agency will be based on the use of nanotechnologies, and agents (i.e., information and robotic nano-agents) will emulate some organs and perform some corrusponding anthropological functions, such as those of physiology, manipulation, and locomotion, intelligence (limited to ratiocination), and interaction with the outside world. The positioning of the anthropic nano-agents actually inside the body of the human being makes it possible to imagine anthropic agents as an important development in which present-day bio-machines (aids, ortheses, and prostheses) will be replaced by meta-machines (anthropic, information, and robotic nano-agents) that together make up an anthropic agency.

The notion of "ubiquitous computing" leads inevitably to the existence of the "ubiquitous computer," which is a computer found anywhere there is the body of a human being and that can be connected without wires and electromagnetically to all computers and robots in the same habitat. By the term "ubiquitous computer," which we may call an anthropic computer, we mean therefore not only portable computers but also future-generation cell phones, in which the two separate information machines unite in one information machine, the "anthropic agent."

The anthropic agent will be hung or worn on the body and will use the most advanced microtechnologies (where the prefix "micro," which means one millionth, signals technologies based on among other things the use of microelectronics, which with devices such as the transistor is able to build objects the size of a micrometer, or a millionth of a meter). We may consider the evolution of the miniaturization of computers and robots as follows:

- *Macrocomputers and macrorobots* are characterized by the traditional technology developed in the 1940s, 1950s, and 1960s, which made it possible to build IT machines much larger than the human body.
- *Microcomputers and microrobots* are characterized by more recent technology developed in the 1970s, 1980s, and 1990s that made it possible to build IT machines of the same size as the human body.
- *Nanocomputers and nanorobots* are characterized by the technology of the near future, which will be developed in the early years of the 21st century and which will enable us to make IT machines that are smaller than the human body.

The functions emulated by nanocomputers and nanorobots, which change in response to different nanoelectronic and nanochemical devices, can be placed in three typologies:

1. *Low-level functions*—those supplied by the independent organs of the body, such as the heart and kidneys, the organs of the five senses (sight, hearing, touch, smell, taste), the muscles, and so on. These functions are activated independently, with decentralized controls mainly at the level of the single organ, even if automatically connected and integrated both with the peripheral nervous system and with the central nervous system (both the brain and the spinal cord).

2. *Medium-level functions*—those supplied by the coordination and cooperation of independent organs of the body, such as the coordination and cooperation observed in walking or in gesticulating. These functions are activated by the peripheral nervous system even if automatically connected to and integrated with the central nervous system.

3. *High-level functions*—those supplied by any part of the functions of the intelligence: computation, deduction, memorization, or intellective functions carried out both in isolation (as in the case of the computer) and in aggregation with the functions of interaction with the physical, "natural," and artificial world, distinct from the body. Here the body is understood as an entity limited to the "natural" component, should the body have on it or inside it artificial organs (prostheses and information devices), as is already happening today and will happen more and more in the future. It is well to remember that interactive functions are carried out both by entering the body (sensory functions corresponding to the perception of phenomena) and by coming out of the body (activating functions corresponding to the production of phenomena).

Therefore, the advent of nanotechnologies, in particular nanoelectronics, represented by spintronics and molecular nanochemistry, will produce, from the metric point of view, devices whose size will be of the order of nanometrics; from the anthropometric point of view, devices whose size, measured in relation to the size of the human body, will be of the order of the size of the organs of the human body; from the anthropotopological point of view, devices that will be placed inside the human body; from the anthropic point of view, devices whose artificial functions will emulate the natural functions of the human body, such as those exercised by the heart, eyes, and ears, and also by the muscular system of the upper and lower limbs, whose purpose is to gesticulate and move about, and by the system of intellective and interactive functions exercised by the central and peripheral nervous system.

Note, therefore, that in the near future we shall see the replacement of the anthropic agent based on microtechnologies, and especially microelectronics, by the anthropic agency based on nanotechnologies, whose agents will be nanocomputers and nanorobots. Therefore the societal agency (distributed and concentrated) offered by ubiquitous computing will not have the ubiquitous computer, that is, the anthropic agent, as a societal agent (or rather, as the agent of the societal agency machine), that is, as the human/machine pole of the human being that belongs to information society. It will have instead an anthropic agency constituted, at a lower hierarchical level than the notion of a societal agent, of anthropic nano-agents.

Still further on, we can intuit that, if an anthropic agent develops more on this lower hierarchical level than on the level of the anthropic agent (i.e., of the whole human being), the anthropic nano-agent itself will be replaced by an anthropic nano-agency, emulating a single organ and function of the human being, in its turn made up of anthropic pico-agents (picocomputers and picorobots).

The last stage in the development could be the replacement of anthropic pico-agents by anthropic pico-agencies, made up of anthropic femto-agents. These in their turn would be made up of groups of molecules of chemical substances (e.g., medicinal), understood as femtocomputers and femtorobots that will unerringly spread in an ordered and guided way (and not in a disordered and erratic way, as happens today with chemical substances contained in medicines) through the whole body to reach the cellular details of the organs that are to be treated with these medicines.

2. ANTHROPIC, DOMOTIC, URBOTIC, AND ORBOTIC AGENCIES

The prospect of the introduction of information and robotic agents in the anthropic agency (the body of the human being), the domotic agency (the house of the human being), the urbotic agency (the city of the human being), and the orbotic agency (the land of the human being) may appear exaggerated, even if their aim is to put themselves at the service of the disabled and able-bodied. However, it is important to emphasize that, among the functions that robots are capable of carrying out today, there are several groups of complex functions.

The first group of complex functions is called conventionally "intelligent functions," where "intelligent" refers to the ability of the robot automatically to resolve a series of problems that it is presented with.

The second group of complex functions consists of "sensory functions," which enable a robot to perceive phenomena that take place both in its own physical structure ane in the environment external to its physical structure. In the former case, the sensors are part of the robot in question and are endowed with the characteristic of self-reference. In the latter case, the robot is able to build a symbolic geometrical model of the environment, and then, thanks to its intelligent functions, use it to construct a coherent path to take. In this case we refer to a *deliberative or symbolic robot*. The robot also reacts to stimuli from the external world, so as to construct behavior that is coherent with its own goals and with the ties deriving from the external environment. In this case we refer to a *reactive or subsymbolic robot*. Because the robot is in constant dialectical contact with events taking place in the outside environment, it is influenced in its behavior and it therefore acts not on the basis of a predetermined and preexisting strategic plan, but on the basis of perceptions expressed in the form of signals. It remembers that a signal corresponds to the value of a physical size that varies in time and space. Because of the difference between the signal and the symbol, with the signal occupying a lower level of abstraction in modeling the phenomena of reality, we use the term *subsymbolic* to describe the information on which these robots work.

A third group of complex functions consists of "autonomous functions." These enable a robot to interact with the human being who imparts commands, not through the imperious communication of simple commands to be carried out in a direct and subordinate fashion, but through the representation of more complex and abstract needs. This leads in synthesis to a problem of service that the robot, whose final goal is to help the human being, is ready to resolve by bringing the operation to its conclusion.

3. ANTHROPOLOGICAL ASPECTS

At this point we may ask ourselves about the possible consequences of the anthropic, domotic, urbotic, and robotic agency for the human being seen as a disabled or ablembodied subject. More precisely, we may ask ourselves if the human being can represent forms of anthropological evolution, given that it integrates with the use of an agency immersed in the "natural" habitat, that is, the body (anthropic agency) and the "natural" and artificial habitat, that is, the home (domotic agency), the land (orbotic agency), or one part of it amalgamated by ubiquitous computing,

It is easy, above all, to state that the so-called "Darwinian evolutionary mechanism," stripped of erroneous "pseudo-biological" aspects, becomes, when focused upon more correctly and precisely, simply but significantly a manifestation of human culture. Machines and artifacts are nothing other than the reflection and consequence of a cultural act: The project and its realization are carried out by an architect, who merges *techne* (art, which in this specific case is planning) with *poiesis* (production, in this specific case the realization). More precisely, considering human culture as the sum of "thoughts thought" by those distinguished people who have contributed more than any others to the creation in the human being of knowledge of reality, we may observe that the architecture of a machine, the "thought thought" of its architect, is a manifestation of human culture. Therefore virtuous selection, which in the evolution of human society leads to the privileging of this or that architectural solution to make a particular machine, arises from the constant simmering, according to a criterion of progressive improvement, of human culture, which is, in the last analysis, knowledge of reality.

Naturally this consideration takes on a more important aspect from the anthropological point of view when it is projected—on the basis of the prospects offered by nanotechnologies such as spintronics and macro-molecular chemistry—onto the vision in the midterm of a possible introduction into the body of the anthropic agent, and in the long term of the possible introductioo inside the body of an anthropic agency constituted of several anthropic agents cooperating with one another. It is evident that the introduction of one or more anthropic agents, that is, nanocomputers and nanorobots, inside the human body does not in any way change the conception of the bipolar human being illustrated earlier, both in terms of the distinction between the human/body pole and the human/machine pole (in this case made up of an anthropic agent or an anthropic agency placed inside the body), and in terms of the unitary nature of the human being, understood as a unitary human/mind subject, who carries out intellective and interactive activities at the two poles.

When we consider the human being subject to the two well-known evolutions that anthropology has described—that is: (a) *genetic evolution*, connected to variations in the chromosome code, which modifies the human being at the biological level, and (b) *cultural evolution*, connected to the influences received from contacts with social networks, with the educational system, with access to mass media and cultural events and instruments, and with the ensuing modifications at the cultural level—then the introduction into the human/body pole of the human/machine pole (or at least a part of it) makes it possible to envisage a third evolution, that is: (c) *technological evolution*, which is connected to transporting artifacts inside the body, made possible by nanotechnologies.

At this point, a further question about the anthropological evolution of the human being, in the light of anthropic, domotic urbotic, and orbotic agencies, is whether the development of the human being is always under the control of the human/mind subject, or whether, on the contrary, this development is triggered by a mechanism of self-alimentation imposed by the human/machine pole, that is to say, an anthropic, domotic, urbotic, and orbotic agency. This issue is connected to the consideration that it is always possible to distinguish between the *operator*—what an operation does—and the *operand*—that which undergoes the operation. At times, in the case of more complex machines, the human being is able to plan machines that, understood as operators, are able to carry out operations on operands that are in their turn also operators that carry out operations on other operands. The pioneering work of Johann Von Neumann on self-replicating automata is a more extreme example, in which the first operator carries out an operation that involves the modification of an operand, which is actually a part of the architecture of the first operator.

A fundamental point in this analysis, which must not be in any way approximate, is therefore the word *part* in the previous sentence. It is not true that the operand, when it is still the operator, changes and evolves autonomously from the human being. What is true is that the human being who planned the "first" operator also defined in its architecture an iterative mechanism that allows the "first" operator its own operand; therefore, if this operand is in its turn a "second" operator, it appears true that a machine, as a "first" operator, can mechanically modify another machine, a "second" operator (which in Von Neumann's case is a "part" of the "first" operator).

The most advanced robots therefore are machines that have been programmed with a "first" program, inserted by the human being, which can construct and modify a "second" program, which involves the specification of
activities that the single robot carries out through interactions with the external
world. The consequence is an inferential modality typical especially of all systems
of artificial intelligence. An example would be a computer programmed to play
chess: This machine carries out a "first" program whose instructions define the
strategic behavior of the artificial player, which constructs a "second" program
whose instructions define the moves of the game that the artificial player plays
against the second player, even if he or she is human.

REFERENCES AND BIBLIOGRAPHY

Somalvico, M. (1984). *Robotica: linee di sviluppo e applicazioni industriali* [Robotics: Lines of de-
    velopment and industrial applications]. Enciclopedia della Scienza e della Tecnica [Encyclo-
    pedia of Science and Technology]. Milan: Mondadori.
Somalvico, M. (1987). *L'Intelligenza artificiale* [Artificial intelligence]. Milan: Rusconi Editore.
Somalvico, M. (1991). *Intelligenza artificiale* [Artificial intelligence]. Enciclopedia Treccani.
    Rome: Aggiornamento.
Somalvico, M. (1997). *Intelligenza artificiale. Aspetti scientifici e applicazioni* [Artificial intelli-
    gence. Scientific aspects and applications]. Atti del Convegno dell'Istituto Lombardo di
    Scienze e Lettere su "L'Intelligenza." Milan: Istituto Lombardo di Scienze e Lettere.
Somalvico, M. (1999). *Intelligenza artificiale*. Atti del Convegno dell'Istituto dell'Enciclopedia
    Italiana su "Corporeità e Pensiero" [Artificial intelligence, Proceedings of the Conference
    of the Institute of the Italina Encyclopedia, "Body and Thought"]. Rome: Editrice Studium.

CHAPTER FIVE

Sade Triumphant:
The Body in Contemporary Art

Raimondo Strassoldo

1. THE BODY IN THE ARTS

All art is a bodily affair. Even in its most spiritual aspects, art is produced by the body and perceived through the body. Traditionally, music has been defined as the most spiritual of arts, because it cannot be touched or seen and because the inner ear is such an inconspicuous part of the body. But the extent of muscular strain implied in its production is well known to anyone who performs it, as well as to anyone who watches a live concert; and music has a peculiar ability automatically to set the body in motion. Written poetry and literature would appear to be even more disembodied; but writing and reading are physical activities, subject to the laws of bodily movements and postures. Chinese poets and medieval copyists competed in the formal elegance of their writings and the implied hand gestures and brush strokes. In our own times, the competition between electronic and paper-based media revolves, among other things, around the different bodily requirements in their use. It is still difficult to read a computer screen in bed, in the toilet, or in a crowded subway.

In the performing arts the body is the main expressive tool. Actors use it along with the voice, and dancers, mime artists, acrobats, and clowns rely almost exclusively on their control of bodily movement. Then there is that new form of art, garment fashion, in which the body exists in symbiosis with the artifact.

In the visual arts, the human body has been by far the most popular subject, at least in the Western tradition. I would guesstimate that, up to the 20th century, 90 percent of sculptures and 75 percent of paintings represented human bodies (the rest being mainly horses and other animals in the case of sculpture, and architecture, landscapes, and again animals in the case of paintings). Architecture is the least corporeal of the arts, although anthropomorphism is a well-known feature of architectural forms, and many other ties with the human body (movement, kinesthesis, functions) can be pointed out.

2. THE NUDE

In Western painting and sculpture, the human body appears in three main genres. The first is the figure (classically, the persona or the mask), which represents a social type, an historical category, or an abstract idea. The figure is identified by signs and symbols, trappings and props—beginning with the dress. The second genre is the portrait, in which subjects are not only located in a precise historical and social situation, but also characterized psychologically in their unique individuality. The portrait is a far less widespread genre than the figure; very few visual cultures (such as the ancient Roman and the modern European ones) have developed it.

The third genre is the nude (and not merely the naked). To qualify as a nude, the representation of a naked body must meet at least two requirements. The first is a certain degree of eroticism. The nude must appeal, however faintly, to sexual desire (Clark, 1956). That is why the crucifix, even though almost naked, cannot be labeled as a nude; nor can pictures of children, cupids, or angels. There are, however, obvious problems of borderline cases and possible improprieties (Higonnet, 1998). The second requirement is that the nude be treated as a "symbolic form" to express universal emotions (not merely love and desire, but also pathos, energy, ecstasy, and so on) rather than sociohistorical categories or individual characters.

The nude was an invention of the Greeks and is a peculiarity of the Western tradition. It flourished in two distinct periods: between the 5th century BC and the 4th century AD; and between the 15th and 19th centuries AD. Only India, of the other high civilizations, had a comparable and, erotically, even more impressive tradition of the nude; although it seems that the Greek stimulus was seminal in developments in India.

The Western nude occurs in two basic models. By far the most enduring is the classical Mediterranean model, first established by the Greeks and revived by Renaissance Italians. Polyclitus spelled out the laws (the canon) of the ideal male body; Praxiteles created the template for the female body. Polyclitus' canon is alive and well, being embodied in most contemporary images of male beauty (Dutton, 1995). The female ideal form, even within the classical tradition, exhibited greater variation in vital statistics, ranging from the plumpness of the baroque to the tautness of the rococo; but Greek marble Venuses, and their imitations in Italian Renaissance paintings, in general still please the eye. The second model is the Gothic northern one, marked by more elongated proportions and fewer curves; it lived on well into the 16th century, and reappeared briefly even later. Since the end of the 19th century, it has been favored by the upper classes and by high fashion.

3. THE BODY IN TWENTIETH-CENTURY WESTERN ART

The climax of the classical tradition of the nude in Western art was arguably reached in the late 19th century, with the highly erotic works of Bouguereau and Cabanel, in the academic style, and of Renoir and Rodin, in the impressionist/expressionist style. In the years around the turn of the century, certain movements, such as symbolism and decadence, developed a taste for a different female body type, sinfully androgynous and emaciated (Strassoldo, 1997).

In the 20th century, Western art split into three main contrasting currents. The first—which survived up to about midcentury—was the traditional art system, centered on the great public institutions for the teaching, conservation, and enjoyment of art (mainly museums, galleries, academies, ministries, and markets). Here the classical nude survived for a while, more or less jadedly.

The second was the avant-garde, based in Paris in the first half of the century and in New York in the second half. With some notable exceptions, the avant-garde ridiculed and eschewed the classical nude, subjecting the human body—like any other object—to all kinds of distortion, dissection, and mutilation. The process began with Degas' and Roualt's unpleasant pictures of prostitutes, and was most clearly manifest in Picasso's *Les Demoiselles d'Avignon*. The representations of the human body usually expressed indignation, rage, hate, and contempt for the present state of affairs (society, nature, the world). Sexuality was usually depicted as base, repulsive, and abnormal, rather than delightful and pleasurable—recall here Schiele's necrophiliac, skeletal, lecherous lepers; the rotting flesh in Kokoschka's and Lowith's bodies; the angular puppets of other expressionists; Dubuffet's public-lavatory-style scribblings; Man Ray's roped and broken Venus; Hopper's lonely naked urban housewives; de Kooning's toothy harpies; the offal clumps of Francis Bacon; Lucian Freud's bloated monsters.

The third area was commercial art. Pictures of the human body and often of the nude were widely circulated by the new cultural industries, starting with the producers of portfolios of "art photographs" (reproductions of nudes taken from high art, or their staged imitations). From its very beginnings, the advertising industry found one of its main tools in the human (mainly but not only male) pleasure in watching beautiful bodies (mainly female). Street posters and illustrated magazines were full of such images (Gabor, 1972). At about the same time, the motion picture industry started to flood the world with animated images of beautiful people, in ever more scanty dress and ever more sexually explicit scenes. A side industry of erotic and pornographic images developed (Webb, 1978). And then of course came the collapse of traditional sexual mores, television, the videocassette, and the Internet. In the second half-century, the infosphere exploded like a supernova of untold billion images of nudes—mostly very beautiful and endowed with supernaturally developed sexual traits.

4. THE AESTHETICS OF THE UGLY

The high-art hostility toward the classical nude can thus be interpreted as a reaction to its banalization in the culture industries. But it can also be set in the wider context of the epochal inversion in the self-definition of art, that is, in the sudden transition from the "fine" to the "ugly" arts, from the aesthetics of beauty to the aesthetic of ugliness (Bodei, 1995).

How did the 20th-century avant-garde justify its strange ways of (mis)treating the human body? A host of rationalizations have been offered:

- Beautiful bodies are rare, or at any rate beauty is transient. In reality, most people are ugly. The images of beautiful bodies permeating academic art are hypocritical, and those in mass culture and commercial art are a misleading sham. True art must represent the truth, that is ugliness.
- All the aesthetic potentialities of the genre have been explored and exploited; nothing new can possibly be extracted by the beautiful body. Because the main duty of the artist is to be original, there is no alternative to the experiments with the ugly body.
- In all times, art has represented ugliness and violence as well: the wicked, the devil, torture, death, the crucifixion, illness, and so on. The beautiful classical nude is only a part of the story. The mistreatment of the human body in art has a long tradition.
- The duty of art is not to convey sensual pleasure through beautiful images, but to irritate, arouse, enrage, displace, and shock; and this can better be achieved through ugliness.
- Capitalist, industrial, bourgeois society is bad, and the artist must denounce it in order to awake the masses and excite them to revolt. Art must represent the darker, uglier side of reality, in order to negate present society.
- In the 20th century, humans have caused the most horrible crimes: war, extermination, genocide, famine. Human bodies have been violently maimed and destroyed wholesale, by the hundred million. In everyday life, bodies are being continually broken by machines, on the roads, and in factories. Experiences and images of such scenes are ubiquitous. Art cannot refrain from representing them.
- After Auschwitz, Hiroshima, and Chernobyl, it is no longer possible to be proud of humanity or to nurture faith in human progress. Humankind is a cancer on the planet, and is rightly doomed to annihilation. Artists, being endowed with greater sensibility and intuition than most people, are only anticipating the inevitable with their apocalyptic visions.

In the course of the first three quarters of the 20th century these doctrines became conventional wisdom, almost clichés, constantly repeated in all brochures for avant-garde art exhibitions and artists' summaries. In the past two or three decades, however, some fresher theories have accompanied the spread of new experiments in avant-garde art:

- Humankind has reached the limits of technical civilization; the only way to survive is to rewind the evolutionary clock and start again from primordial conditions. Humankind must return to a state of nature, to the authenticity of the tribal way of life, with all the implied dirt and violence, including all sorts of manipulations of the body typical of primitive cultures.

- Humankind has reached the threshold of an evolutionary leap comparable in magnitude only to what happened with the hominid revolution itself, some million years ago, and with the invention of symbolic culture, about 50,000 years ago. Humankind has now assembled a sociotechnical system that can evolve by self-design, according to its own inner rules and goals, toward a future of unlimited potential (omniscience and omnipotence). We are entering a posthuman world, in which the individual, the person, the subject, the body is an increasingly irrelevant part of the system. The natural, biological endowments of individuals are less and less adequate to the requirements of the sociotechnical system. Humans can and must be reprogrammed genetically, redesigned, improved, equipped with new mechanical and electronic prostheses (cyborgs). The process will inevitably entail some experimental failures, mistakes, and suffering; and some monstrosities may occur. However, that is the road to the future, and artists should cooperate with scientists and technologists in making it come true (Bell & Kennedy, 2000; Lunenfeld, 2000).

5. BODY ART: A TYPOLOGY

As remarked at the outset, all art is basically body art. But only in the 1960s was the term *body art* coined to refer to a growing variety of peculiar practices stemming from what were originally visual artists (Vergine, 2000). The main idea was to de-reify art, to decouple artistic production from the practices of industry, market, and merchandising. Artists were to refrain from producing objects that could be sold and bought, and instead create ideas, visions, awareness, emotions, concepts. The artist must evolve into prophet, theorist, philosopher, ideologist.

Some of the early names of these practices were conceptual art, behavioral art, actionism, performances, happenings. Their enaction closely resembled what was done by the more established performing arts: Conceptual artists did something (e.g., cutting themselves and drawing blood, or masturbating, or quartering lambs, or washing linen, or planting trees) in front of a live audience. Usually, however, they had the performance recorded by photography, film, or video camera, and thus converted into something more akin to the traditional visual arts. They often used props, tools, and costumes, but the main medium of the performance was the artist's body. Only the setting (art galleries, ateliers, exhibitions, even open spaces, instead of theaters) and the life history of the artist marked the difference between this kind of body art and the more established performing arts.

But the term body art is also used for other artistic practices:

- The exhibition of living human or animal bodies, in their more or less natural state.

- The use of human bodies as brushes or knives—Yves Klein smeared his models with blue paint, and then threw and pressed them on the canvas.
- The emphasis on the body gestures by which the art object is produced. Jackson Pollock had himself photographed while moving around and on the canvas and dripping his pigments. Reference to his postures, gestures, and muscularity was considered an important symbolic content of his canvas.
- The body as canvas. The painting of the surface of the body is an ancient and widespread cultural practice. In tribal communities it has a great variety of symbolic functions. In modern societies it is mainly limited to women, and has basically cosmetic and sexual functions. Hair stylists and beauticians have always done it. Body artists have used it as part of their performances or in its own right.
- Permanent alterations of the body: tattooing, piercing, branding, scarification, constraints, cosmetic surgery. Many cultures have treated the human body as an object to be manipulated for symbolic and aesthetic purposes. The inclusion of tattooing in this category, instead of the preceding one, is because of its permanent and three-dimensional (in the thickness of the skin) character. Piercing, branding, scarification, and constraints have for a long time been confined to non-Western cultures, but neo-primitive, neo-tribal, and neo-ethnic lifestyle trends have made them popular again in postmodern Western society, and artists have explored their aesthetic potentialities—usually on their own bodies because it is not easy to find subjects willing to undergo permanent bodily modification just to satisfy someone else's creativity. Some artists in California are reported to be experimenting in the alteration of the interior of the body, for instance blood composition, by ritually inoculating each other with the HIV virus.
- Visual exploration of hitherto neglected aspects of the body. Macrophotography has been used to produce large images of tiny parts of the external anatomy, such as wrinkles, pimples, and hairs. Microvideocams have been inserted by artists (e.g., Mona Hartum) in various bodily orifices in order to benefit the public (often interactively) with visions of the pulsating interiors of their bowels.
- The coupling or integration of the human body with technological systems (techno body art). Bio-engineering has been working for generations in this field, for both medical and ergonomic purposes. More recently, science fiction and cyberpunk culture have elaborated at length on the theme, envisioning a cornucopia of integrated human/machine organisms, prosthetic people, cyborgs, androids, and so on. But, as L. Borgese once remarked apropos of architects, in circus parades the acrobats are followed by the clowns; some artists (e.g., Stelarc and Antunez) have also explored these perspectives, designing and using technological prostheses in their performances. The possibility of aesthetic interaction, via technological communication systems, between bodies widely separated in space, is one of the

frontiers of "media art." The pupils of the Köln School for Media Art have developed an electronic system of videocams, computers, and robots that allows two faraway individuals to engage interactively in sadomasochistic erotic practices.

- The use of dead human bodies, or parts thereof, as materials for art. This practice is not new. Many primitive cultures used parts of dead bodies (shrunken heads, skulls, and so on) for personal, home, and public decoration. Egyptian mummies and medieval Christian relics were certainly treated with an eye for aesthetics, although the main reason was religious. In modern Western society, the use of dead human body parts is strictly reserved to medical and scientific practices. However, there are borderline cases. In Germany a few years ago, great success was enjoyed by an exhibition of artistically modeled and posed skinned human corpses, which a couple of medical doctors had managed to stabilize and crystallize chemically so as to give the muscle tissues the appearance of crisp freshness. In London, a fashionable young sculptor found a way to coat chunks of human bodies electrolytically in metal. When caught he was arrested, because artists, unlike medical researchers, do not (any longer, or not yet?) have the right to use dead human bodies as raw material.[1] Is that an acceptable ethical limitation on artistic creativity?

6. DE SADE IN THE ARTS

In the contemporary art system, exhibitions focusing on the human body are very popular and successful.[2] In most of them, most of the items are not pretty in the conventional sense. The visitor is presented with homely, ugly, or monstrous bodies, or with humiliated, suffering, defaced, punctured, deformed, opened, disassembled, penetrated, tortured, tied, reconstructed bodies, or with body/machine contraptions.

The texts written around such events (in catalogs, brochures, panels, press releases, reviews) usually refer to some of the clichés summarized earlier in Section 4. I would not deny that most of them have some basis in these rationalizations; but I would add a simple, encompassing explanation. Such exhibitions are staged, and people flock to them, (a) because sadism is a basic human drive, and (b) because we are living in a culture heavily imbued with Sadean thought.

What is to be seen in those exhibitions has an undeniable family resemblance to what the Marquis de Sade wrote down 200 years ago, in many thousands of

---

[1] We have not been able to retrieve the clippings from the *Corriere della Sera* arts pages with more detailed information on these cases. A picture from the German case is displayed on the cover of Perniola (1998).

[2] One can randomly recall the "L'ame au corps" at the Grand Palais in Paris (1993), the Venice Biennale on the theme of human identity (1995), the "Flesh Factor" exhibition at the Linz Ares Electronica Center (1997), "Der Anagrammatische Koerper" at the KMZ in Karlsruhe (2000), the "Mennesket" at the Arken (Denmark) Museum of Modern Art (2000).

pages. His heroes and heroines delight in destroying beauty (having abused it sexually), and also in having sex with ugly, deformed, decrepit, revoltingly sick, and crippled bodies. They use living bodies as pieces of furniture, cut them up and reassemble them in fanciful ways, compose *tableaux vivants* with bodies of various types and in various stages of manipulation, decorate the venues of their feats with corpses and dismembered body parts. They exalt in the design and use of the most sophisticated machines for delivering torture and death. They enjoy altering living (unanesthetized) bodies by all sorts of surgical operations: skinning, ripping open, sewing, chopping, grafting. Victims are hung in midair by hooks and straps. De Sade's heroes and heroines deal erotically with body exuviae and excreta, and, of course, are enthusiastic cannibals (one of the most memorable is an ogre named Minski, just like the computer science guru). They also enjoy wholesale massacres: Ferdinand of Naples has hundreds of victims theatrically butchered at each of his parties, and the French prime minister, the Prince of Saint Fond, a typical Sadean hero, during one of his orgies details the plans for the extermination (by poisoning water supplies) of 5 million surplus Frenchmen.

The Marquis' immense literary output[3] describes all sorts of horrors that can be performed on the human body; hardly anything is left to the imagination. His is the definitive catalog in this genre. His imitators (e.g., Swinburne, Apollinaire, Lautréamont, and de Mandragues) pale beside him.

The place of the Marquis in Western culture is peculiar. On the one hand, the attempt was made to suppress him: He was locked up in jail and the madhouse for most of his adult life; and his writings were outlawed and destroyed. But they seeped out, and circulated underground like a karstic stream. There were almost no references to his work in official culture for nearly a century, until the science of psychiatry used his name—along with that of Masoch—to label a mental sexual disorder. On the other hand, there is little doubt that his influence on all Western culture has been enormous. True, de Sade himself can be seen partly as the result of a peculiar mix of Enlightenment *forma mentis* (his endless tracts attempting to demonstrate the soundness of his philosophy rationally and scientifically) and of Romantic content (hatred of God, nature, reality). But it can also be argued that much subsequent Romantic culture, from painting (Fuseli, Delacroix, Gericault) to literature (Baudelaire) to philosophy (Nietzsche), was imbued with Sadean visions. The influence was especially strong, of course, in the Parisian intellectual milieu, from where it radiated to the whole of the Europe. A more detailed analysis of Sadean motifs in high culture would be interesting, but impossible in the present context. The list of de Sade's open or covert disciples would be impressively long.[4] Instead, mention

---

[3]See Lely (1952/1957) for a bibliography. A bibliography of 65 items (some of them more than a thousand pages long) has been compiled by Seaver and Wainhouse (1966). A much more limited one can be found in Englisch (1927), which however is commendable for its extensive review of erotic and pornographic French literature in de Sade's times.

[4]Two cases in point: Gabriele d'Annunzio, one of the most important Italian authors at the turn of the century, whose works are full of Sadean motifs; and Michel Foucault, perhaps the most famous Parisian (and hence global) intellectual of the 1960s and 1970s, whose philosophical interests as well as private life are clearly Sadean (see Miller, 1993).

should be made of the presence of Sadean motifs in the developing mass culture: For the past quarter-century, popular novels and newspapers have been filled with a mix of sex and death.

In the 19th century, de Sade was the *deus absconditus* in high culture; after Freud, Lombroso, and Krafft-Ebing, his name came out of the closet (and the libraries' infernos), and references to his name gained legitimacy, first in scientific discourse and then in general culture. It can be argued that most avant-garde authors of the 20th century were well acquainted with de Sade, and many admitted it openly. Guillaume Apollinaire, the prophet of Cubism, besides writing a book on de Sade tried to imitate him in the *Les 10.000 vierges*. Kafka called de Sade the patron saint of the 20th century. However, in my opinion, the explicit acknowledgment of de Sade's influence and the open study of his works and thought represent only a tiny fraction of his de facto presence in our culture. This is so deep, so pervasive, and of such immense scope as to blind us. As is well known, the strongest ideologies are those that we are unaware of.

Only in the past half-century he has been the object of extensive scholarly analysis.[5] Some of the studies seem to find some redeeming social value in his writings, as manifestos of absolute individual freedom. In my opinion, there is no way to consider him as anything other than by far the most uncompromising destroyer of all human values, the greatest prophet of universal hate and cosmic annihilation, the complete anti-Christ.

In the second half of the 20th century, the Sadean mixture of eros and thanatos, of pleasure and pain, of sex and violence, of orgasm and horror, became the main fare of mass culture too. People look eagerly for it in all media, which compete in supplying ever greater and more extreme doses of what they want. The popularity of art exhibitions featuring the bloody body can be interpreted, at least to some extent, as a side effect of this general trend.

7. CONCLUSION

The Marquis de Sade's objectives in his relentless literary output are plain enough: to excite himself in the contemplation of his terrifying fantasies, and to convince the reader of the soundness of his ideas about the fundamentally evil essence of the world. For him, the only natural imperative is to pursue one's own pleasure; beauty gives the more intense pleasure by being destroyed, and pleasure derives mainly from the suffering and death of others. His heroes state again and again that all the horrors they perpetrate have the sole function of "making my sperm squirt a shade more hotly."

What then are the objectives of those who produce Sadean images in contemporary Western culture? One, clearly, is to make money. The cultural industries

---

[5]Among the best known are Bataille (1957), de Beauvoir (1951/1952), Blanchot (1949), Brega (1962), Horkheimer and Adorno (1947), Klossowski (1947), Paulhan (1946), and Zolla (1964). See also Paglia (1990).

are in business for profit, and they discovered a long time ago that sadism sells very well and that there is an unlimited market for it.

But a string of further questions comes to mind. Why do people like to read and watch such stuff? Does symbolic/artistic/commercial sadism perform other (perhaps "redeeming") social or moral functions? Are there innate limits on the extension and intensity of sadism in our culture, or should limits be imposed by law?

As to the first question, a number of answers can be aired. The taste for sadism can be imputed to the manipulation ("corruption") of individuals by the cultural industries themselves, in a self-amplifying vicious circle. It can be imputed to general conditions in industrial society: alienation from nature and community, the violence of mechanical systems, violence against nature, a surplus of psychophysical energy no longer used in work or war, the lack of familiarity with real sex, violence, and death in everyday life, the banalization of "normal" sex, and so on. It can be imputed to society in general: the long prevalence of war, male domination. Personally, I am inclined to think that the contiguity of pleasure and pain, of aggression and excitement, of eros and thanatos, is a basic constituent of human nature, linked with our evolution as predatory primates, as killer apes, as carnivorous and cannibalistic hunters. It is the deep, dark side of our psyche; the evil in our genes, which most social systems have tried to harness; our original sin, from which religious and ethical systems have tried to redeem us through repression and sublimation, and that now is simply let loose.

As to the second question, the debate has been long and inconclusive. After myriads of psychological and sociological studies, the old two theses are still unresolved. According to one, images of sex and violence tend to excite people to imitation; they are one of the main causes of sadistic behavior. According to the other, those images have a cathartic function; they allow sadistic tensions to be released in the virtual, symbolic, or fantastic realms, thus keeping them clear of behavior. Of course, these two main theses can be endlessly refined, for example, in relation to the different effects according to the stage of psychic development.

Finally, it is hard to imagine that the cultural industries—of which the art system has largely become a segment—would autonomously impose limits on the quantity and quality of sadism in their outputs. In Western society, art and culture, as well as science, have for two centuries developed on the principle of "no limits": No social, moral, or political restrictions can be imposed on them. Thus, all previous criteria of "obscenity" in the representation of sex and violence have been rapidly transgressed, with the one exception, so far (for how long?), of pedophilia.

Whether this trend can last indefinitely is open to question. Many thinkers suggest that cultural systems sway from one extreme to the other, and that eras of hedonism give way to eras of asceticism. Malraux predicted that "the 21st century will be spiritual, or will not be." Personally I think that the sociotechnological conditions in which we live are so radically different from what humankind has experienced so far that no known theory can help us. What the future holds for relationships between the body, the aesthetic realm of the arts, science and technology, and ethics is wholly unclear, and rather hair-raising.

REFERENCES AND BIBLIOGRAPHY

Alfano Miglietti, F. (1997). *Identità mutanti. Dalla piega alla piaga: esseri delle contaminazioni contemporanee* [Changing identities. From the fold to the sore: Beings of contemporary contaminations]. Ancona/Milan: Costa & Nolan.

Bataille, G. (1957). *L'Erotisme* [Eroticism]. Paris: Minuit.

Bell, D., & Kennedy, B. M. (Eds.). (2000). *The cybercultures reader*. London: Routledge.

Blanchot, M. (1949). *Lautréamont et Sade* [Lautréamont and Sade]. Paris: Minuit.

Bodei, R. (1995). *Le forme del bello* [Forms of the beautiful]. Bologna: Il Mulino.

Brega, G. P. (1962). Introduzione [Introduction]. In *De Sade, Opere scelte*. Milan: Feltrinelli.

Clark, K. (1956). *The nude. A study in ideal form*. New York: Doubleday.

Comar, I. P. (1999). *The human body. Image and emotion*. London: Thames & Hudson.

de Beauvoir, S. (1951/1952). Faut-il bruler Sade? [Must we burn de Sade?]. *Temps modernes* [Modern times]. December 1951 and January 1952.

Dijkstra, B. (1986). *Idols of perversity. Fantasies of feminine evil in fin-de-siecle culture*. New York: Oxford University Press.

Dubin, S. C. (1992). *Arresting images. Impolitic art and uncivil actions*. London: Routledge.

Dutton, K. D. (1995). *The perfectible body. The western ideal of physical development*. London: Cassell.

Englisch, P. (1927). *Geschichte der erotischen Literatur* [History of erotic literature]. Stuttgart, Germany: Puttmann Verlag.

Gabor, M. (1972). *A modest history of the pin-up*. London: Pan Books.

Higonnet, A. (1998). *Pictures of innocence. The history and crisis of ideal childhood*. London: Thames & Hudson.

Horkheimer, M., & Adorno, T. W. (1947). *Dialektik der Aufklarung* [Dialectic of enlightenment]. Amsterdam: Querido.

Klossowski, P. (1947). *Sade mon prochain* [My arrival to Sade]. Paris: Seuil.

Lely, G. (1952/1957). *La vie du Marquis de Sade* [The life of Marquis de Sade]. (2 vols.). Paris: Gallimard.

Lunenfeld, P. (2000). *Snap to grid. A user's guide to digital arts, media, and cultures*. Cambridge, MA: MIT Press.

Miller, J. (1993). *The passion of Michel Foucault*. New York: Simon & Schuster.

Paglia, C. (1990). *Sexual personae*. New Haven, CT: Yale University Press.

Paulhan, J. (1946). Le Marquis de Sade et son accomplices [The marquis de Sade and the accomplices]. Preface to *Les infortunes de la vertu* [Preface to the 1946 edition of the Misfortunes of Virtue]. Paris: Edition du Point du Jour.

Perniola, M. (1998). *Disgusti. Le nuove tendenze estetiche* [Misfortunes. The new aesthetic tendencies]. Genoa/Milan: Costa & Nolan.

Seaver, R., & Wainhouse, A. (Eds.). (1966). *The Marquis de Sade: Three complete novels, Justine, Philosophy in the Bedroom, Eugenie de Franval and other writings*. New York: Grove Press.

Strassoldo, R. (1997). La moda, l'arte e l'idealtipo longilineo [Fashion, art and the longilineal ideal type]. In L. Bovone (Ed.), *Mode*. Milan: Angeli.

Strassoldo, R. (1998). *La natura nell'arte: il corpo e il paesaggio* [The nature of art: Body and landscape]. Audiovisual CD-ROM, University of Udine, Udine, Italy.

Vergine, L. (2000). *Body art e storie simili. Il corpo come linguaggio* [Body art and similar stories. The body as language]. Milan: Skira.

Warwick, A., & Cavallaro, D. (1998). *Fashioning the frame. Boundaries, dress and the body*. Oxford, England: Berg.

Webb, P. (1978). *The erotic arts*. London: Secker & Warburg. (Original work published 1975)

Zolla, E. (1964). Prefazione [Preface to de Sade]. In *De Sade, Le opere*. Milan: Longanesi.

CHAPTER SIX

The Narrated Body:
The Representation of Corporeality
in Contemporary Literature

Ada Neiger

Huxley maintains that, in the past, literature too had its Manichaeans; writers, that is, who described humans' rationality and spirit and ignored their physical reality on the grounds that it was it vulgar. Description of the body was restricted to certain parts of the face: the eyes, whose expressive eloquence was emphasized, the lips, the source of visual messages, and the hair, also able to emit signals. For many writers the human body was comparable to a continent, of which they explored few territories. Only pornographers and certain burlesque artists concerned themselves with the buttocks, genitals, and other intimate areas usually carefully concealed from public gaze.

Now that the taboo on the body has been lifted, contemporary authors have made it the object of reflection: It may assume multiple functions, it is the object and subject of desire, and it is an instrument of communication. Moreover, it can be made up, tattooed, disguised, manipulated, mutilated, and fitted with high-tech "prostheses."

Female writers of the second half of the 20th century, with bashfulness abandoned and no longer forced to adopt a prudish reserve, spoke freely about their own bodies and those of others. Thus Alda Merini candidly described her "belly bloated by disease and the malice of life ... her stomach so swollen that it suggested long and difficult pregnancies" (Merini, 1994, p. 54) and all the changes that the passage of time, illness, and experience of mental hospital had produced in her. Here, as elsewhere, Merini does not complain about her appearance, nor does she resent the many electric shocks that have burnt "entire segments" ("interi segmenti") of her brain (Merini, 1994, p. 65), because when she sits down to write she can always rely on her intestines, her bowels.[1] Merini's absence of hypocritical prudishness prompts a number of general reflections.

We may begin with Max Scheler (1986), who analyzed the concept of nudity and clothing and reached the conclusion that humans do not wear clothes solely to protect themselves against the weather. Clothed or naked, the body emits signals. Skillfully disguised, masked, veiled, or stripped naked, it arouses a broad array of emotions and sentiments, ranging from pleasure to revulsion. The body's language is crucially complementary to verbal communication, and, unlike the language of rationality, it is simple to understand. As Dacia Maraini (1996) pointed out, it is young women especially who use the language of the body, which "triggers male desire with such ease ... and which draws its strength from two thousand years of history and from the female body's age-old ability to 'say without saying'" (pp. 66, 67).[2]

Numerous writers in literature have described the languages of seductive females, given that the body of a woman is almost invariably treated as "an object to be looked at and possessed,"[3] whereas a man's has other prerogatives—it is "desiring and therefore, in becoming, ready for action and for transformation" (Maraini, 1996, p. 40).[4] In short, the female body is akin to an object, a commodity, or a consumption good. Women consequently take pains to make their appearance pleasing and attractive. Although examples of the enticement of women's bodies abound in literature, scant space has been devoted to female ugliness. A significant example was provided by Malaparte (1981), who in *Pelle* dwelled on the "midgets" ("nanerottole") of Naples, "bald and toothless [who] crawl on all fours, yelping and foaming at the mouth: they resemble the freaks of Breughel or Bosch" (1981: 23).[5] These repulsive women are flanked by a rabble of others: the "ghastly toothless women, with their withered faces encrusted with cosmetics" (p. 101),[6] who expose the half-naked bodies of their children to the gaze of the Moroccan soldiers, their possible purchasers as slaves; the "proletarian effeminates with their curly black hair, red lips, dark and shining eyes" (p. 76);[7] the women sitting on the Chiaia Steps peddling their bodies by splaying their legs, "disgustingly to display their black pubes amid the glistening pink of naked flesh" (p. 46).[8] There are other horrors besides, because the novel abounds with descriptions of bodies, sometimes detailed with disconcerting realism.

---

[1]"Having been deprived of my natural brain by the electric shocks that burnt whole segments of it, in order to write I interrogate my intestines, my bowels" (Merini, 1994, p. 65).

[2]"che mette in moto con tanta semplicità la macchina del desiderio maschile ... [e che] appoggia le sue forze su duemila anni di storia e certamente comporta una antica sapienza del 'dire senza dire' che appartiene al corpo femminile."

[3]"oggetto da guardare e da possedere."

[4]"desiderante e quindi in divenire, pronto all'azione e alla trasformazione."

[5]"calve e sdentate [che] si trascinano carponi, mugolando e sbavando: paiono mostriciattoli di Breughel o di Bosch."

[6]"le orribili donne sdentate, dal viso scarno e vizzo incrostato di belletto."

[7]"efebi proletarii dai capelli ricciuti e nerissimi, dalle labbra rosse, dagli occhi scuri e lucenti."

[8]"in modo orribile, mostrando il nero pube fra il roseo bagliore della carne nuda."

Malaparte (1981) wrote that the Neapolitans had fought against dying during the war; then, when the allies liberated their city, they had fought to live. After struggling with dignity during the war, they had committed every wickedness to survive when liberated. Contrasting with Malaparte's colourful prose is Primo Levi's laconic style, and the more engaged descriptions of another memorialist, Elie Wiesel, who provided us with a lucid portrait of humanity trampled underfoot: the prisoners in the Nazi extermination camps.

Deformed bodies call to mind an entire strand of literature dealing with persons afflicted by physical disabilities. Authors with handicapped kin have drawn on the experience to produce works whose protagonists are afflicted with severe physical disabilities.

Clara Sereni, for instance, presented a small gallery of young disabled people in her short stories collected in *Manicomio Primavera* (1990). Giuseppe Pontiggia's recently published *Nati due volte* uses the story of a spastic boy to reflect on physical illness. Paolo, the main character in the novel, is able to move only with great difficulty:

> He walks close to the walls of the houses so that he can have something to lean against if he stumbles. His gait is clumsy. Rather than following his body's commands it seems to exploit its weight as he plunges forward in sudden spurts.... Those who see him for the first time stop and stare. He notices them. I have the impression that as he hobbles along he grimaces with suffering. But perhaps that is not so. Perhaps he is only taking care not to fall over. He is used to being watched; it's me who cannot resign myself. I grimace with pain, and it is this that unites us, at a distance. (Pontiggia, 2000, pp. 231, 232)[9]

"To the disabled who struggle not to become normal but to become themselves" (p. 41)[10] runs the epigraph to Pontiggia's novel. This is an invitation to examine the concept of normality and to accept those different from ourselves. To the question "who is normal?," Pontiggia unhesitatingly replied "no one," wisely adding, "when one is faced by diversity, the first reaction is not to accept it but to deny it. And one does so first by denying normality" (p. 41).[11] But normality does not exist. Pontiggia not only pointed out our errors but, like the majority of contemporary writers, suggested ways to rid ourselves of prejudices and stereotypes. In discussing what he called the "need for the normal" ("bisogno della norma"), which he likened to "a virus made invulnerable by the medicine taken to kill it,"[12] he offered

---

[9]"Cammina lungo i muri delle case, per avere un appoggio, se incespica. L'andatura è sgraziata e, anziché seguire i comandi del corpo, sembra sfruttarne il peso, precipitandolo talora in avanti con accelerazioni improvvise.... Chi lo vede per la prima volta spesso non se ne accontenta. Si ferma e si volta a guardarlo. Lui se ne accorge e ho l'impressione che arranchi con una smorfia di sofferenza. Ma forse non è così, lui bada solo a non cadere, è abituato a essere osservato, sono io che non mi rassegno. Ho una smorfia di sofferenza ed è quello che ci unisce, a distanza."

[10]"Ai disabili che lottano non per diventare normali ma se stessi."

[11]"quando si è feriti dalla diversità, la prima reazione non è di accettarla, ma di negarla. E lo si fa cominciando a negare la normalità."

[12]"un virus reso invulnerabile dalle cure per sopprimerlo."

the following illuminating insight: "It is not by denying differences that we combat it, but by changing the image of normality" (p. 42).[13]

Flanking Pontiggia is Clara Sereni, whose upbeat invitation is not to lose hope even when we are assailed by apparently irremediable pain. Of particular significance is her intense short story "Tutto si impara," in which without overblown rhetoric she described the birth of a handicapped baby and the dismay of the mother, who felt "a pang of disgust at her uncontrollable body, at this piece of herself ripped from within, at the blood, sweat and shit that splashed from her with life" (Sereni, 1990, pp. 25–26).[14] The dismay continues when she is presented with her new-born baby, "red, wrinkled and ugly, the gaping voracious mouth ... as weird as if it came from another planet: its large and elongated head, its ill-assorted body" (p. 26).[15] With an effort, the mother clasped the baby to her body, "so that she might no longer see it, but their bodies recognized each other and they exchanged warmth. She breathed deeply, trying to catch the rhythm that they'd taught her during the course—'You'll learn to love it,' they'd told her—everything can be learnt" (p. 27).[16]

Barbara Garlaschelli (2000) introduced us to Davì, who does not suffer from a disability but sports an eccentric hairstyle, in a short story scattered with valuable observations that help us to understand others. Garlaschelli invited us not to dwell on the boy's bizarre appearance, but rather to imitate the young red-headed woman who is struck, not by Davì's crest of green hair, but by the way he walks, which reflects his state of mind: "While everyone else was agitated, running and walking as fast as they could, he seemed to move in slow motion. He strolled as if he had no particular destination in mind but rather a pleasure to cultivate" (p. 70).[17] Davì is an emblematic character, who attracts attention with his punk outfit, his head weighed down by pointless earrings, his weird hairstyle, and his hair dyed green. But the anonymous red-haired woman, bewildered by "voices saying nothing and producing only noise" (p. 70)[18] and irritated by the mass of drab passers-by with mobile phones glued to their ears, "all of them the same, all wearing the same clothes, marching in unison towards who knows where" (p. 69),[19] when she looks out of her window and sees this comical boy with a yellow flower in his hand, she is not shocked at the sight. Rather the small figure glimpsed from her office window has almost a therapeutic effect upon her. It calms and soothes her.

---

[13]"Non è negando le differenze che lo si combatte, ma modificando l'immagine della normalità."

[14]"trafittura di rifiuto: per il proprio corpo incontrollabile, per quel pezzo di sè che le si strappava di dentro, per il sangue il sudore e la merda che schizzavano via da lei insieme alla vita."

[15]"rosso grinzoso e brutto, la bocca aperta e vorace ... E strano, come venuto da un altro pianeta: la testa grande e allungata, il corpo disarmonico, incompatibile."

[16]"così non lo vedeva più, i corpi però si riconoscevano e si scambiavano calore. Respirò profondamente, cercando il ritmo che le avevano insegnato durante il corso:—Imparerò ad amarlo, si disse.—Tutto si impara."

[17]"Fra tanti che si agitavano e correvano e camminavano spediti, lui pareva procedere al rallentatore. Passeggiava come uno che non ha una meta da raggiungere ma un piacere da coltivare."

[18]"voci che non dicevano nulla e producevano solo rumore."

[19]"persone tutte uguali, vestite tutte uguali, che marciavano compatte verso chissà dove."

Old age, too, can be likened to a handicap. The elderly are unproductive and they consume little. Society finds them unpleasant and pushes them aside. Yet there are writers who devote space to the elderly, highlighting their weakness and decay but also emphasizing their undiminished need for affection and their continuing ability to abandon themselves to amorous passion. The stories of senile love recounted by Barbara Alberti (1991) and Elena Gianini Belotti (1995) provide a powerful antidote to the sexual ostracism that society reserves for the elderly. One is reminded in this regard of certain thoughts of Simone Weil. "The human body is the scales for which the supernatural and nature are the counterweights," she wrote (Quaderni, III, p. 315; quoted in Obinu, 1989, p. 95). "This condition of modern life everywhere upsets the balance between the spirit and the body.... The civilization in which we live, in all of its aspects, crushes the human body. The spirit and the body have become extraneous to each other" (quoted in Obinu, 1989, p. 72).

Pier Paolo Pasolini (1976) warned young people against what he called the "rhetoric of ugliness" ("retorica della bruttezza"; p. 63) in an article significantly entitled "Siamo belli, dunque deturpiamoci" ("We're beautiful, so let's disfigure ourselves"):

> For a number of years young people have done everything to make themselves look ugly. They adopt hideous hairstyles. They are not content unless they are completely masked and disfigured. They are ashamed of their curls, of the ruddy glow of their cheeks, they are ashamed of the light in their eyes due to the candor of youth, they are ashamed of the beauty of their bodies. Those who triumph amid this madness are the ugly ones: those who have become the models of fashion and behavior. (1976, p. 63)[20]

There is another way to disfigure oneself: the method used by the anorexic, the bulimic, or the obese, the people of every age prey to eating disorders. These are usually women whose clinical histories show markedly maniacal behavior. Fragile creatures unable to accept themselves, they tend to establish clinging relationships in friendship and love. Externally they appear strong, but their intimate friends and families, aware of their conviction that they are insufficiently loved, that they are rejected or even abandoned, fear their reactions and often avoid them lest they be overwhelmed. Anorexia nervosa was widespread in the 20th century, and in the 19th the art of fasting aroused great interest among the general public and scientists. Kafka's short story, "Ein Hungerkünstler" (1996), describes the artists of hunger, and several women writers have described self-starvation, usually autobiographically (e.g., Fusini, 1996).

---

[20]"Da alcuni anni i giovani, i ragazzi fanno di tutto per apparire brutti. Si conciano in modo orribile. Fin che non sono del tutto mascherati o deturpati, non sono contenti. Si vergognano dei loro eventuali ricci, del roseo o bruno splendore delle loro gote, si vergognano della luce dei loro occhi, dovuta appunto al candore della giovinezza, si vergognano della bellezza del loro corpo. Chi trionfa in tutta questa follia sono appunto i brutti: che sono divenuti i campioni della moda e del comportamento."

People can shape their bodies by fasting, dieting, or bingeing.[21] However, they are not always able to change them as they wish, so they resort to science or technology. In the case of a transsexual trapped in a male body who claims the right to alter his body and to assume an identity that matches his feminine sensitivity, recourse to surgery is almost obligatory. The existential hardships suffered by a Brazilian transsexual and his metamorphosis brought about by surgery and hormone treatment have inspired Fabrizio de André's musical composition, *Princesa*, and Maurizio Jannelli has written a book about the story of Fernanda Farias de Albuquerque.

Mention of the story of the Brazilian transsexual invites us to reflect on science and technology. Science sets out to study nature and identifies the laws that govern it. Technology exploits the discoveries of science to subjugate nature. With the proliferation of scientific discoveries and technical inventions, the myth has arisen of the!machine able to boost the performance of the human body.

Interest in mechanical instruments dates back to the 17th century. Since then much progress has been achieved, and advances in technology have revolutionized the manner in which we work, use our leisure, and even conduct our lives. The body is the instrument with which humans relate to the environment, and every society develops the "body techniques" by which, Mauss tells us, the individual learns through imitation and training. Tools were invented as extensions of the body that compensate for the imperfections of the human organs. Their place was then taken by machines that, according to Mariella Combi (2000), "externalize the body and its potential as objects that interpose themselves between man and nature" (p. 100).[22]

The car is a machine that has not only satisfied the individual's need for mobility but produced radical changes in human existence. J. G. Ballard (1995) wrote in the introduction to his novel *Crash*, in which he interwove catastrophic traffic accidents, that he has used the car both as a sexual symbol and as a metaphor for the human condition in contemporary society (p. 6). According to Ballard, *Crash* is the first pornographic novel based on technology, and its principal purpose is to warn the reader of the pandemic cataclysm distinctive of modern industrial societies (pp. xi, xii).

We are distracted from Ballard's gloomy and violent scenario by Giuseppe O. Longo, who drew our attention to less disquieting horizons. His "Vicini" (1999) is an enjoyable *divertissement* with a surprising *noir* denouement. The story is reminiscent of Buzzati, with its interweaving of fantastical elements and an underlying anguished intolerance. In the world described by Longo, citizens line up at government offices to have mobile phones implanted between their ears and jaws. They are not dissuaded by the fact that the implantation requires three days in hos-

---

[21]On the historical cultural roots of anorexia, see the interesting book by Vandereycken and van Det (1995).

[22]"prevede l'esteriorizzazione del corpo e delle sue potenzialità, è un oggetto che si interpone tra l'uomo e la natura."

pital and is costly, or by the fact that the implant is impossible to remove afterward. Contrary to expectations, the prosthesis does nothing to improve communication and understanding among people. The new "biotelephonic technology" produces only chatter, gossip, suspicion, hostility, and irritating noise. Sebastian, the man who refuses to desecrate his body by accepting the machine into his flesh, is the maverick who, contrary to all expectations, is not overwhelmed by the "conformists."

Less catastrophic than Ballard and less ironic than Longo is Gibson, whose writings focus on a future peopled by cyborgs. The bodies of these hybrid creatures are a blend of flesh and inorganic technology. These mutants carry forward a cyberculture based no longer on territorial nationalism (as has been rightly pointed out), or on territorial belongings, or on institutional relations, or on power relations, but on aggregation around shared loci of interest—games playing, the sharing of knowledge, cooperative learning, and open collaboration. This tendency toward virtual communities is wedded to an ideal of deterritorialized, transverse, and free human relationships (Lévy, 1997).

Eschewing futuristic speculations, a 1997 novel by the Italian writer Carmen Covito describes the emergence in everyday life of virtual realities and the transformation of the body. It does so with subtle irony, without irritating sermonizing and without upsetting us with the disastrous consequences of technological supremacy. The title itself of the book is encouraging—*Benvenuti in questo ambiente* ("Welcome to this world")—although the contamination of elements (organic and inorganic) and the virtual community could easily give rise to malaise and identity problems.

One of the characters in the book who engage our imagination is Sandrina, so coldly rational that her mother describes her as a woman who "has a calculator instead of a heart" (Covito, 1997, p. 31).[23] A young career-woman, Sandrina is, again according to her mother, "the technological genius of the family" (p. 8).[24] Her elder brother Ugo is a well-known plastic surgeon who has performed numerous operations, including a sex change and complicated "surgery in assisted bodily reconstruction" (p. 252)[25] performed on his mother, who volunteered as a guinea pig for her son's experiments. With her body now monstrously disfigured, she lives in isolation in the basement of her villa, using a computer to communicate with the outside world.

Besides the surgically mutilated body of Lucia, there is the tattooed body of the transsexual and the ring-pierced body of Annapaola. Fully six passages in the novel describe Annapaola's piercing (Covito, 1997, pp. 80, 85, 90, 104–105, 248). The girl's nose sports "two silver hoops threaded through different holes but so close together that they tinkle when she laughs or when she snorts" (p. 88).[26] Her

---

[23]"al posto del cuore ha una calcolatrice."

[24]"il genio tecnologico di famiglia è lei."

[25]"interventi di rielaborazione corporea assistita."

[26]"due cerchietti d'argento infilati in due buchi diversi ma talmente vicini che possono anche urtarsi tra loro quando lei ride forte o, come adesso, sbuffa con sufficienza."

nipples are pierced and so too is her navel, and "even down below, where her shaved ash-blonde down formed an inverted triangle, there glinted further semi-hidden rings" (p. 104).[27] Annapaola exposes her body to the gaze of Nureddin, a young immigrant, who is embarrassed at the sight of so much metal attached to intimate and delicate parts, and is unable to understand what might drive a woman to thread rings through her flesh (p. 105).

Another memorable character in the novel is Marco. We follow him during a journey in his Mercedes, when he notices a "purring intimacy between the car's mechanical organs and his mind, which still perceived the boundaries of his body and the machine with a vague detachment ... that was more exciting than a sexual encounter.... Metal and flesh were welded together. And their driving force was indissoluble" (Covito, 1997, p. 216).[28] Mention should also be made of the pigeon caught by Ugo and Sandrina with the intention of transforming it into a bionic robot (pp. 223, 224, 237).

This rapid survey of the representation of corporeality in Italian fiction thus closes with Covito's novel, a work that I believe fully corresponds to what David Cronenberg (1997) described as the meaning of art today. It is, in fact, a book in which the author follows her phantasms, which "may be uncomfortable for some, but I believe that art resides precisely where minds are perturbed."[29]

REFERENCES AND BIBLIOGRAPHY

Alberti, B. (1991). *Delirio* [Raving]. Milan: Mondadori.

Ballard, J. G. (1995). *Crash*. London: Vintage.

Belotti, E. G. (1995). *Adagio un poco mosso* [Adagio a little moved]. Milan: Feltrinelli.

Combi, M. (2000). *Corpo e tecnologie. Rappresentazioni e immaginari* [Body and technology: Representation and imagination]. Rome: Meltemi.

Covito, C. (1997). *Benvenuti in questo ambiente* [Welcome in these surroundings]. Milan: Bompiani.

Cronenberg, D. (1997). Interview by Francesca Alfano Miglietti in *Virus*, no. 10, January.

Fusini, N. (1996). *La bocca più di tutto mi piaceva* [The mouth was the thing I liked more]. Rome: Donzelli.

Garlaschelli, B. (2000). *Davì*. San Dorligo della Valle, Italy: Edizioni EL.

Kafka, F. (1996). Ein Hungerkünstler [A hunger artist]. In *Erzählungen*. Munich: Martus Verlag.

Lévy, P. (1997). *Cyberculture. Rapport au Conseil de l'Europe* [Cyberculture. Report to the Council of Europe]. Paris: Éditions Odile Jacob.

Longo, G. O. (1999). Vicini [Neighbors]. In P. Rinaldi (Ed.), *Il galateo del telfonino* [The book of etiquette of the cellular phone]. Faenza: Mobydick.

Malaparte, C. (1981). *La pelle* [The skin]. Milan: Mondadori.

Maraini, D. (1996). *Un clandestino a bordo* [A clandestine on board]. Milan: Rizzoli.

---

[27]"perfino laggiù, dove una rada peluria biondo cenere disegnava un triangolo rovesciato, brillavano altri anelli seminascosti."

[28]"ronfante intimità tra gli organi meccanici e la mente, che ancora percepisce i confini del corpo e della macchina come un vago distacco ... è più eccitante di uno scambio sessuale.... Metallo e carne sono ben saldati. E la coppia motrice è indissolubile."

[29]"possono essere scomodi per molti, ma credo che l'arte sia proprio dove si turbano le coscienze."

Merini, A. (1994). *Reato di vita autobiografia e poesia* [Crime of life, autobiography and poetry]. (Luisella Veroli, Eds.). Milan: Melusine.

Obinu, S. (1989). *I dilemmi del corpo. Materia e corporeità negli scritti di Simone Weil* [The dilemmas of the body. Material and corporality as shown in the writings of Simone Weil]. Poggibonsi: Lalli.

Pasolini, P. P. (1976). *Lettere luterane* [Luterane letters]. Turin, Italy: Einaudi.

Pontiggia, G. (2000). *Nati due volte* [Born twice]. Milan: Mondadori.

Scheler, M. (1986). *Über Scham und Schamgefühl* [Shame and feelings of modesty: Posthumous writings]. In *Schriften aus dem Nachlass, Band 1*. Bonn: Bouvier Verlag Herbert Grundmann.

Sereni, C. (1990). *Manicomio Primavera* [Spring Asylum]. Florence: Giunti.

Vandereycken, W., & van Det, R. (1995). *Dalle sante ascetiche alle ragazze anoressiche. Il rifiuto del cibo nella storia* [From the ascetic saints to the anorexic girls. The rejection of food in history]. Milan: Raffaello Cortina.

CHAPTER SEVEN

Real People, Artificial Bodies

Leopoldina Fortunati

The human body has always endowed itself with artificial limbs and other devices to extend and multiply its capabilities in both a cognitive and an operative sense (Longo, 1999). But, now, information and communication technologies (ICTs) have not only come near to the body but are also entering it, for the time being in the form of microchips, but in time also in the form of implants or genome modifications to grow the peripheral devices that we desire. ICTs were brought into effective service under the umbrella of medicine, not just at a preventive level, but also diagnostically and therapeutically (Katz, 2003). In Great Britain, experimentation then also insinuated itself into the domestic sphere in order to facilitate command over household appliances and any mechanism capable of being worked electronically. In the United States, ICTs were experimented with to facilitate the finding of missing persons. The shift was therefore from the therapeutic field to quality-of-life engineering.

Medicine is the platform across which this approach to the body and its penetration by various technologies has passed. The process was very simple: Because the end (prevention, diagnosis, and therapy) justifies the means (penetration of the body), the progressive invasion of the body itself was allowed, as was the movement toward replacing all its various parts, first with artificial limbs and then with organs transplanted from living bodies, or even removed from corpses, or developed in the laboratory.

In this chapter I attempt to reconstruct the premises and one of the many fundamental processes that have made it possible for ICTs to penetrate the human body. The structural premises of the penetration of the human body by technologies are given, I believe, by the facts that:

- The boundaries of the body are very uncertain.
- The body has always been imagined as a machine, in the sense of both reducing the human body to a machine, and creating a machine similar to a human being.

The process that I have decided to analyze in its various articulations is the further separation of mind from body produced as a result of the development of artificial intelligence. I examine in particular three tendencies that increase the alienation of the body: (a) the mind setting itself up as the sadistic pole, and the body as the masochistic pole; (b) concentration of the mind in the male, and of the body in the female; and (c) development of the mind in the West, and of the body in the so-called Third World.

1. THE BOUNDARIES OF THE BODY

The first structural premise that has played a role in the social acceptability of the transgression of the boundaries of the body is their indeterminacy. To state that the body is enclosed in "topological" limits, indicated by the skin, is—from a communicative and operative point of view—arbitrary and substantially inexact (Longo, 1999). Given that the voice is an extension of the body, we could say that the body reaches where the voice does, that is, a good many meters away from the body itself. Even Montaigne had something to say about the voice as a boundary zone of the body, and posed the problem of its nature. If the body belongs to the single individual, to whom do the words that come out of the body belong? Montaigne's conclusion was that the words belong half to the person who utters them and half to the person who listens to them.

Apart from the voice, the definition of the extent of the bodily "I" includes other elements too. Personal space, for example, is the zone that immediately surrounds the individual, and that, seen as a projection of the "I" in space, tends to increase the extent of the body (Sommer, 1967). Its defense, personalization, and control are indispensable not only for the expression of one's personal autonomy but also for the definition of modes of social interaction.

Clothing is another element that contributes to extending the bodily "I." By increasing the apparent size of the body, it gives the impression of a greater bodily extent (Lotze, 1897; cited in Flugel, 1992, p. 46): When we are clothed, we occupy more space. Through the effect of "confluence," wrote Flugel, "the enlargement of the human figure, in reality due to clothes, is unconsciously attributed to the body that wears them, because that is the most vital and interesting part of the whole" (p. 98).

2. THE PUBLIC IMAGINATION AND THE BODY AS MACHINE, AND THE CREATION OF MACHINES SIMILAR TO HUMAN BEINGS

Another important premise that has permitted and favored the penetration of technologies into the body is that the human imagination has through the ages created a particular family of monstrous figures. In the most ancient structures of the imagination, which right from primordial times reflected the fact that the body was the first and most immediate technological tool, we find archetypal figures of the body as machine: the automaton, golem, android, and robot (Fortunati, 1995). These figures however represent not only the common experience of the body as a fundamental technological tool, but also the specific envy that men have always felt for the capacity of women to give life. The automaton, the golem, the android, and the robot are incarnations of the male dream of using available scientific knowledge to create a human being.

In the history of humankind, these figures are a recurrent theme in various forms of narration, but they have also been the focus of continuous experimentation. For centuries humankind has experimented—even in concrete form—with ways of building a machine that will emulate a human being and a human being that is similar to a machine.[1] The practice started with the attempt to build machines that could imitate human beings. These machines were called "automatons" (*thaumata*). The technology of automatons, prized by Aristotle because of their capacity to create intellectual surprise and stimulate theoretical questioning, developed rapidly in the Hellenistic period, and especially in the Alexandrian period, in the ancient science of Pneumatics (Ferrari, 1984, p. 225 ff.).

In more modern times, we have gone from the automaton to the robot. However, this shift did not represent further development in the automaton, but was a leap, a break in the process of construction of the machine/human being. The leap was illustrated particularly well by Baudrillard (1990, pp. 64–66) as an attempt to pass from the mechanical and theatrical counterfeit of the human being that was the automaton to a technical principle that is able to produce the equivalent of a human being. If the automaton has to astonish and entertain in attempting to be more natural than the human being, the robot, thanks to its similarity to human beings, needs only to demonstrate its mechanical effectiveness in comparison with their capacity for work.

The latest chapter in the history of the machines that attempt to replicate human beings is artificial intelligence. Artificial intelligence has tried to construct a mind without a body, that is, an undisturbed intelligence that will imitate the "higher" functions of the biological mind, avoiding any interaction with an environment considered a source of distraction (Longo, 1999).

In practice, however, there has been no lack of applications in the opposite direction as well, that is, in producing a human being similar to a machine. The idea of the living being/machine took on great importance with the development of experimental technology, especially for the construction of artificial limbs (Zanca, 1991, p. 33). The French surgeon Paré (1509–1590), driven by the necessity of supplying artificial limbs for mutilated soldiers, developed arms and hands with complicated mechanisms. The project of dismantling the body/machine was to be carried on by many researchers from the second half of the 17th century onwards, until it arrived at the present-day implant engineering of pacemakers, and so on.

---

[1]This circular relationship between human beings and machines we owe to Herasistratus, the great Alexandrian anatomist, who was the first to build a marvelous automaton.

In conclusion, our imagination has not found it particularly difficult to assimilate the process of penetration of technologies into the body, because for thousands of years it has been the domain for experimentation. Moreover, imagination and action have always been very closely intertwined in humankind's attempts to make the body match the machine, and vice versa. So today the penetration of technologies inside the body is actually less shocking than we might have imagined.

3. THE FATE OF THE MIND WITHOUT A BODY AND THE BODY WITH A DISTORTED MIND

The advancing penetration of technologies inside the body was possible because artificial intelligence had pushed the separation of mind and body toward a point of no return. This process, despite its notable results on the scientific plane in various disciplines, has ended up by weakening both the mind and body, as we see next.

The separation of the mind from the body has distant origins and developed at various crucial moments in the history of the West, for instance the birth of the technology of writing. It is the literate society, in fact, wrote Ong (1977, 1982), that experiences the separation of subject and object. When the unity of nonliterate societies breaks down, intellectual appropriation of an object can occur only through our separation, and thus our distancing, from it. In this way, abstraction is created, and knowledge becomes analytic. Every kind of experience is fragmented into uniform units, becoming a function of the efficient management of both distance from and intimacy with the world and hence also the body (Ong, 1977, 1982).

Following on from this (and thanks also to Christianity), in Western culture the management of distance from the body became a clear-cut separation and antagonism, the primary form of every dialectal opposition. The Cartesian conception of the body as *res extensa*, as a machine, was also important in this development; it was the premise that triggered the process of devaluation of the body that is today being radicalized to the full. And the Marxist analysis of alienation shows how the capitalist system has made individuals extraneous to themselves, reducing them to mere containers of the capacity for work, and thus forcing them to bow to the commands of those who hold the means of production.

Today, however, the accelerated separation of the mind from the body induced by the development of artificial intelligence is accompanied by an inextricable knot of processes. As I said before, I here analyze three of these: (a) positing the mind as the sadistic pole and the body as the masochistic one; (b) concentration of the mind in the male, and of the body in the female; and (c) the development of the mind in the West, and of the body in the so-called Third World.

The Mind as the Sadistic Pole, the Body as the Masochistic Pole.   The loss of the mind/body unity that leads to the construction of a disembodied mind, the progressive devaluation of the body, and the accentuation of the alienation of the individual not only is being amplified but is undergoing a radical change with the development of artificial intelligence. With what effect?

The most immediate consequence is that the body ends up being treated with destructive fury. Despite many apparent indications that the body is being cared for—gyms, fitness regimes, diets—the body is in fact being subjected to real suffering (if not actual torture—e.g., self-inflicted wounding, piercing, tattoos, plastic surgery) and, on another level, forms of therapeutic rage. This attitude to the body reflects the fact that current modes of social relations tend to acquire concreteness and realness inside the body and implode inside it, with the result that individuals' extraneousness to themselves assumes dimensions that are often psychologically disturbing.

Let us try to reconstruct this implosion in broad outline, observing it from Foucault's viewpoint of the microphysics of power. My thesis is that sadomasochism, understood not as a perversion but as a basic model of social relations seen as power relations, affects individuals in their deepest structural self. Sadomasochism has imploded inside individuals, leaving a mark on the relationship between their mind and their body.

To describe this process, let us examine an emblematic power relationship, that between man and woman. At a social level this relationship has always been formally presented as comprising a sadistic pole (man) and a masochistic one (woman). In his essays *The Economic Problem of Masochism* (1924), Freud (1978, p. 7) spoke of a particular masochism in women, and considered this tendency a drive, as a way for female nature to express itself. I, in contrast, think this tendency on the part of women is a historical result, not an inclination of female nature.

In any case, the sadomasochistic man / woman relationship has always been at risk of being overturned informally in the domestic sphere, where the roles can be reversed (the man becoming masochistic and the woman sadistic). In fact, on a general level, individuals have probably always achieved a balanced fulfillment of sadistic and masochistic roles, seeing that, as Contri wrote (1983, p. 165), humankind has never resolved the problem of the black hole between impotence and arrogance. And it is perhaps the social dimension of behavior, as well as the extremely complex quality of networks of relations, that supplies us with an important key to reading the economy of the drives: According to Freud, it is not by chance that we are dealing with mixtures of life and death impulses.

There remains, however, the fact that this tolerance of pain is incomprehensible in itself. Freud too defined pain, understood not as the signal for flight but as a goal, as an enigma in respect to the economy of the life of drives. The question that at present underlies this enigma is why does the libido today, at a mass level, not manage to expel the drive to death or destruction from the body, toward the exterior? One plausible answer is that in a global society the social elements that generally facilitate expulsion are missing. It may be interesting to indicate at least some of them:

- A lack of paternal power, caused by the decline in authoritarianism and by the crisis of authority—this has led to a reactivation of the Oedipus complex at a social level.
- A lack of a political counterpower, of mass organization and collective projects for changing society—the crisis in political parties, trade unions, and movements has left the problem of political strategy unresolved, which is where drives of destruction, appropriation and will for power used to be concentrated, on the exterior.
- A great transformation of power and a hiding of its workings, so that it is now becoming divorced from property, money, and politics.

So a specific question arises: what sense of guilt in the collective unconscious, what social need for punishment, drives us to seek pleasure in pain? Freud perhaps would urge us to think of the space in the unconscious that does not differentiate between pleasure and pain and that is capable of guiding every action, obviously with different conscious motivations. Lacan, on the other hand, would perhaps eliminate the issue, saying that there is a pain drive in relation to the body and its boundaries (cited in Contri, 1983, pp. 170–171). I prefer to think that the real answer to the aforementioned question is the lack of firmness that these generations had from their parents, and the great unresolved problem of primary and secondary education.

Today, therefore, sadomasochism is introjected into the individual (and increasingly expressed in sexuality) through the breaching of the mind/body unity. As a result, the mind sets itself up as the sadistic pole and the body as the masochistic one. It is in fact the mind, inscribed in that body, that "rationally" programs the body's sufferings, that exhibits an increasingly boundless dominion over it and shows up its slavery. It is the mind that cultivates the distorted idea that the body can be packaged at will, seeing it as raw material whose shape is more or less modifiable. It is therefore always this mind that programs recourse to plastic surgery to make the face younger, to enlarge or reduce the breasts, to implant developed biceps, to reduce thighs, to shape the hips, buttocks, belly, and so on.

It is the same mind that is overvalued at a social level, whereas the body is undervalued, or at most valued at an instrumental level. The mind, in the form of artificial intelligence, has shown that it can live without the body and allows itself the right to do what it likes with the body. This attitude of impatience with and contempt toward the body is increasingly made clear and represented materially. Since the punks, we are less and less reluctant to exhibit the body suffering "creatively." Tattoos, piercing, self-inflicted wounding—all these signs speak of the suffering of the body. But they have little to do with the symbolic significance that they used to have. For the "primitive" they were a demonstration of courage or beauty, within a "warm" mode of thinking that did not recognize a separation between thinking subject and object, and therefore did not perceive a separation between mind and body (Lévi-Strauss, 1962). For the citizen of modernity such symbols are the "cold" expression of an ornamentation that has cost pain. Its value in fact resides not so much in the artistic quality of the sign as in the exhibition of the amount of pain that it cost. This representation of the sadomasochistic relation in the mind/body of the individual points above all to the fact that this relation has been internalized, that it is involved and therefore expresses a regressive valence. But at the same time it points to the fact that the ideological sticky mass that covers and distorts the structure and meaning of social relations no longer "holds," and is showing wide cracks in its social acceptance.

This suffering is represented in concrete terms in the body. But the body does not speak just for itself; it speaks, more broadly, for the whole of nature and its sufferings. In the absence of a strong political pact with civil society, we are facing a similar kind of violence if we consider the inconsistencies and conflicts that the development of science is producing and that economic development is generating as regards nature. If this is the case, what is so sensational or so scandalous about the sadomasochistic use of technologies in a body already programmed to represent suffering, already made dependent on masochistic pleasure?

The Mind in the Male, the Body in the Female. When they penetrated the body, technologies encountered no lines of defense or trenches, but a defenseless body, if only because it was still widely associated with the female. The separation of mind and body followed the same pattern as the traditional separation of culture and nature, in which historically the male gender was attracted to the first pole and the female to the second; so the present opposition between mind and body has attracted the male to the mind and the female to the body. The male, having the power of elaboration, representation, and planning of the world, has more to do with linear thought and intellectual work, whereas the female is more and more confined to speaking the language of the body. Despite the often important incursions of women into the sphere of education, the labor market, and politics, women continue to be assigned, often excessively, to the body.

This is another aspect of the implosion inside the individual of the modes in which the present structure of social relations is presented. That is, the man/woman relationship is imploded inside the individual, probably signaling an important moment in the move toward the androgynous. For a long time psychology has taught us that every human being has both male and female components, even if in men the male part is dominant and in women the female. But the phase that we are going through is forcing us all—men and women—to develop the *logos* (i.e., male thought based on rationality and logic) and not the *metis* (female thought founded on intuition and astuteness). The thinking of globalization is linear, not circular. And at the same time, we are pushed to express a female attitude of care and attraction in relation to the body. Male power and economic sufficiency do not automatically make the male body beautiful any longer. In fact there is a growing male presence among consumers of cosmetic products, beauty treatments, operations, and so on. The man increasingly has to spend money and dedicate time to take care of his body and to make himself attractive.

Given that everyone's mind has to incline itself to the male and everyone's body to the female, every individual has to be simultaneously more male and more female. We will all, therefore, have an increasingly undifferentiated sexual characterization. Sexual identity is already very uncertain and shifting today—and sexual practices are less regimented along precise lines. This new leap in the social construction of sexually less differentiated individuals does not however translate into changes in the sexual division of labor, which still appears very rigid. Men, even if they are more female, are decidedly reluctant to take on the burden of half the domestic labor and the work of care, because this would mean a lengthening of their work day. For this reason the hidden opposition continues to be confirmed between work that produces goods, ideas, services, and strategies and work that produces and reproduces individuals; between work that is mainly male and paid and work that is still too female and unpaid; in other words, between value and lack of value.

This new order of masculinity and femininity, however, will have as a consequence not the valorization of the woman become more male, but the devaluation of the man become more feminine. Men's resistance to doing their "share" in the home is not enough to prevent themselves being devalued, because their new relationship with the body is already a way of devaluing them, and because they continue to be reproduced in a technologically underdeveloped sphere. In fact, reproductive labor continues to be hardly affected by artificial intelligence, which is not able to resolve any of the difficult problems of domestic labor. Despite the promises often offered (Maldonado, 1997, p. 60), the effect on the house of technologies connected with artificial intelligence has in fact been seen in the production of instruments that are in reality little more than gadgets (the intelligent refrigerator, the smart house, etc.). And it is also in this insignificant form of gadgets that communicative technologies have so far approached the body. Perhaps it is this marginal or superfluous importance of theirs that is responsible for the failed appearance of antibodies necessary to defend the integrity of the body.

The Mind in the West, the Body in the Third World. In post-Fordist society, the penetration of the body by technologies was able to take place without any great protests, because the body had been demoted to a mere work tool and had therefore become increasingly a "worker." At the same time, the mind, which is where the power of decision and command is concentrated, became more and more the "boss." What imploded here inside the individual is therefore the capitalist relation, the traditional class relation of bosses and workers. All human beings have in fact become simultaneously bosses and workers, because their mind has a "boss" relation with the body treated as a "worker." On top of the old division between manual and intellectual work a more subtle division is being created, between content—the capacity to work—and container—the body. This new division of labor, apart from strengthening the division between manual and intellectual work, is strengthening the separation of the mind from the body, placing them in opposition. This paradoxical situation can only maximize the alienation of individuals.

But what can a worker do without class consciousness? Marx understood that a worker class, to be that, in other words to become a subject capable of reformulating the social order, must not only refer to its technical colocation inside the production cycle, but also acquire a political consciousness. Unfortunately, the body has not so far expressed political consciousness or, rather, individuals have not become aware that it is on the body, indeed on the lost unity of mind/body, that a really crucial political operation is being played out.

The question is still more complicated by the fact that, at the same time, the dislocation of manual labor far from intellectual work has been favored, that is, the dislocation of production "elsewhere" in respect to planning and management. What has happened therefore is that more and more intellectual work, creation, ideation, research, strategy, and command remain in the First World, whereas the "factory" (production processes, sometimes integrated with information technology [IT] systems) has been moved, often to the Third World, in a relation that is still one of subordination. The mind is concentrated in the West, the body in the Third World. The dislocation, in this case geographical too, of the body serves to keep the body devalued. The mind in the First World, even if it has lost close contact with its corporeal extension, has however gained possession of space in the sky, in the air, as a place of communication and information. This vertical expansion of the West has been favored by the volatile aspect of the word. The extra space that has been gained is of occupation and dominion, because information serves to reduce uncertainty, more through the narcotic weight of its publicity than through the semantic or cognitive weight of its message. The Third World has remained instead more horizontal. It is enough, however, to belong to a place, to be born, in the case of the West perhaps as mind, but in the case of the Third World certainly as body.

In this context, the mind states its identity more and more in the singular (a single network), whereas the body's identity is increasingly plural. Indeed, it is becoming the more and more numerous bodies of workers, who share the condition of being subordinated, controlled, and commanded and subjected to a single categorical imperative: obedience to a global or, rather, cosmic order. We are dealing with bodies without minds, and therefore in demand for material labor. It is obvious that the First World, precisely because it absorbs and is increasingly made up of intellectual work, is becoming rarefied and ethereal, increasingly mind separated from the body. The Third World, in contrast, which houses production processes and much manual labor, is more and more bodies and gross matter. I must however stress that the venture of the mind separated from the body, even in opposition to it, means on the one hand that the mind without a body is condemned to develop a limited and distorted intelligence, and on the other that the body without a mind must be satisfied with a pseudointelligence.

If this analysis is correct, it means that the body is today the place *par excellence* of technological innovation and political struggle, and that, if a common reorganization can be made, it will be around the body, or better still around the body / mind unit. But, to answer my initial question about how we have arrived at the penetration of the body by communication and information technologies, all we need to ask is: How could a body without a political consciousness ever have withstood its penetration by technologies?

4. CONCLUSIONS

In this chapter I have tried to analyze why the human body, which is the emblem of naturalness, has let itself be penetrated, without any great outcry, by technologies. It just remains to add that the coming near to and penetration of the body by communication and information technologies can be seen as the expression of a great male domination in the political and economic management of social processes. This has a clear ancestry of the *logos* and the archetypal figures of the imagination, ranging from the golem to robots and automatons. Therapeutic machines and means of communication will probably serve to open the way to constructing a body reduced to a machine. Individuals will be inserted in an IT architecture of the everyday that will make them act in tune with other machines of a very generalized automation process. At least three issues remain, however, that need further investigation:

1. The implosion of the power relationship (sadomasochism) inside the individual indicates that there is a use for technologies that respond to this logic. Are we facing a crucial phase of social reorganization?

2. The assimilation of machines inside the body follows the pleasure principle (the pleasure of well-being, the recovery of health, etc.), but includes a pact with the death wish. Does the grafting of the inorganic onto the living prefigure a stable fusion with the state of death?

3. Centuries after *habeas corpus*, globalization is going beyond the exploitation of the labor force as work capacity, to command and control over the entire body reduced to a machine, and above all, being exercised as a continuous cycle. Is the implosion of the capitalist relationship inside the individual a way of removing significance from this milestone of Western history?

REFERENCES AND BIBLIOGRAPHY

Baudrillard, J. (1990). *Lo scambio simbolico e la morte* [Symbolic exchange and death]. Milan: Feltrinelli. (Original work published as *L'échange symbolique et la mort*. Paris: Gallimard, 1976)
Contri, G. B. (1983). *Tolleranza del dolore* [Supporting pain]. Milan: Shakespeare.

Ferrari, G. A. (1984). Meccanica "allargata" [The wider mechanics]. In G. Giammatoni & M. Vegetti (Eds.), *La scienza ellenistica* [The Hellenistic science] (pp. 19–47). Naples: Bibliopolis.

Flugel, J. C. (1992). *Psicologia dell'abbigliamento* [Psychology of clothing] (9th ed.). Milan: Angeli. (Original work published as *The psychology of clothes*. London: Hogarth Press, 1930)

Fortunati, L. (1995). *I mostri nell'immaginario* [Monsters of the imagination]. Milan: Angeli.

Freud, S. (1978). *Il problema economico del masochismo* [The economic problem of masochism]. In *Opere, 1924–1929* [Works, 1924–1929] (Vol. 10). Turin: Borignhieri. (Original work published 1924)

Katz, J. E. (Ed.). (2003). *Machines that become us: The social context of personal communication technology*. New Brunswick, NJ: Transaction Publishers.

Lévi-Strauss, C. (1962). *Le pensée sauvage* [Savage thought]. Paris: Librairie Plon.

Longo, O. G. (1999). *Il nuovo Golem. Come il computer cambia la nostra cultura* [The new Golem. How the computer is changing our culture]. Rome / Bari: Laterza.

Longo, O. G. (2001). *Homo technologicus* [Man, the technological species]. Rome: Meltemi.

Maldonado, T. (1997). *Critica della ragione informatica* [Critique of IT reason]. Milan: Feltrinelli.

Ong, W. J. (1977). *Interfaces of the word*. Ithaca, NY: Cornell University Press.

Ong, W. J. (1982). *Orality and literacy. The technologizing of the word*. London: Methueo.

Sommer, R. (1967). Small group ecology. *Psychological Bulletin, 67*, 145–152.

Zanca, A. (1991). Il mondo degli automi tra manierismo e secolo dei lumi [The world of automata in mannerism and illuminism]. In U. Artioli & F. Bartoli (Eds.), *Il mito dell'automa* [The myth of the automaton] (pp. 31–39). Florence: Artificio.

PART II

The Body Communicating Between Technology, Fashion, and Identity

CHAPTER EIGHT

Cross-Cultural Comparisons of ICTs

James E. Katz, Mark Aakhus, Hyo Dong Kim, and Martha Turner

1. BACKGROUND

In our semantic differential study of the style and function dimensions of mobile telephones, we sought to generate and analyze data that would help us explore generally the conceptual underpinning of the mobile phone from a user's viewpoint, and more specifically some of the style and fashion aspects of perceptions about the mobile phone.

Although fashion issues may seem distant from the traditional topics of the communication discipline, we think it is an arena in which communication research holds considerable promise. For instance, fashion as a "second skin" reflects presumptions about communicative settings while also projecting a sense of how others should engage one's self. Thus, the affordances[1] designed into clothing facilitate and inhibit forms of human interaction, as in the design of uniforms for NASA (National Aeronautics and Space Administration) astronauts (Dominoni, 2003). This is an example of an expert designer trying to anticipate the best style of clothing so that it becomes one with the wearer—that is, it fits her or his needs and persona. In a word, it works.

But, there is another sense of fashion that places the ordinary person in the center of browsing and choosing among possibilities and, in a sense, identities, thus re-creating and reinventing the resources in the situation to organize themselves in the pursuit of everyday action. (And here we borrow some insights from the domestication model.) We are interested in how people fashion or put together their communicative presence through the accouterments of communication and perpetual contact. Thus, the meaning of various technologies and their placement in one's personal communication and information infrastructure (Katz & Aakhus, 2002) are central to everyday people's sense of fashion. For example, in this con-

---

[1]The term *affordances* refers to the perceived and actual properties of the object, primarily those fundamental properties that determine how the object would likely be used.

text Aakhus (2001) pointed out how the individualized design of communication and information technologies works against the ability of people to collaborate in managing their collective goods and resources while facilitating their ability to avoid others to whom they have obligations. We investigate here how people judge the meaning of their mobile phone relative to that of other devices and the ancillary question of whether these judgments are just local, situated judgments or are representative of a more general, implicit gestalt about the role of personal communication devices in everyday life.

The line of attack was launched in light of the mushrooming use of information and communication technologies (ICTs). Among them, few if any technologies have spread as quickly throughout the world as the mobile phone. There have been enormous changes in patterns of public and private behavior, as well as in personal and corporate investment. These transformative technologies have been accompanied by extremely modest interest from the professional social science world. The current study is part of this as yet small literature. Its purpose is to understand better the fit and role of ICTs in daily life. We seek this understanding not only because we find ICTs' often surprising uses worthy of investigation in their own right, but because a better understanding can offer important insights into humans' relationship with their built environment.

Of course there has been interest in this topic for some time, including the masterful work of Ogden and Fischer. Yet, within the broad field of academic interest, we can, at the cost of losing important nuances, discern major trends or themes of investigation to position our theoretical and empirical foci.

Functionalist Approaches.   The first theme is the diffusion of innovations, often identified with Everett Rogers and vigorously pursued by many others, including Ronald Rice (Rice & Webster, 2002). This line of thinking looks at diffusion curves for various innovations, and identifies individual psychological traits and personal network configurations that foster or impede the spread of new technology throughout a social system.

This broad functionalist approach was complemented by the "needs and gratifications" approach. In general, research in this tradition examines what people articulate as communication needs and then investigates the extent to which their consumption patterns map onto these needs. Two collateral efforts have also been undertaken as the development of the "needs and gratifications" model matured.

Structuration.   The domestication model, pioneered by Silverstone and Haddon (see Haddon, 2003), describes how ICTs were encountered by families and eventually integrated into household production routines. Playing on the husbandry notion implicit in the term, they saw these technologies as originally existing in an untamed or wild state, and then being tamed and brought into the household, rather like pets. Just as many pets enrich and are part of domestic life, so too the advocates of this perspective saw the same process occurring in terms of ICTs.

On the other side of the ledger, there was growing interest in workplace uses of ICTs. Here the works of Orlikowski (1992) and Poole and DeSanctis (1992) are important. They focus on the use of these technologies in a business setting. The essential factor in this line of research is how the norms and conventions of everyday work-oriented behavior shape the function of workplace technology.

Looking at these phenomena from a meta-level, we see that, in effect, three tines of a fork are used to attack the problem of the human adoption, use, and understanding of ICTs. These strands may be characterized as the diffusion of innovation model (in the center), and household and organizational use on either side. All are materialist-based models, though we also hasten to add that the models are sophisticated and sensitive to the value aspects of human behavior. Uses and gratifications approaches have not been primarily concerned with meaning and interpretation. Diffusion approaches focus on interpreting and predicting broad patterns of adoption. The structuration approaches focus on the gap or mismatch between the technological design and norms for organizing (e.g., households and workplaces). Each approach provides an important vantage point; however, a broad area of exploration and explanation remains. How people incorporate personal communication and information technologies into their everyday action and their understanding of everyday action have not received much attention, and some research has suggested the importance of adopting such a view of ICTs (Aakhus, 2001; Dudley, Steinfield, Kraut, & Katz, 1993; Frissen, 2000).

We by no means intend to fill all the lacunae we see in the current state of the field through our own study. Rather we merely want to contribute some understanding of the fashion and identity issues of personal ICTs in the hopes that we can get a more heuristic insight into the phenomena. Thus, we examine the following questions: What meaning do mobile phones hold for people relative to other personal technologies? Do these meanings vary by national cultural background?

Answers to these questions will help us better describe the relationship between people and their personal technologies, and thus contribute to understanding issues of fashion and design. For instance, if people see their phones as expensive and more comparable to Walkmans than to televisions, this can directly illuminate how they may display the device and what situations might moderate considerations of appropriate use. But these practical questions bring us back to the larger theoretical issues. We did not presume that the personal technologies hold a functional relationship. We do not appeal to uses and gratifications because there is no way in principle to distinguish the relationship between a use and a gratification.

The particular project reported here was undertaken within the larger context of advanced analysis of ICTs and their social ramifications, which is a central focus of the Department of Communication at Rutgers University. Already some significant publications have resulted from this effort (Katz, 2003; Katz & Aakhus, 2002), and we expect that more work will be conducted in the years ahead.

2. METHOD

In spring 2000, a research group at Rutgers University conducted a series of focus groups. They constructed a series of semantic differential questions based on issues that emerged from the discussions and from topics considered of theoretical interest, based on prior research. In this line of thinking, we were inspired by the earlier semantic differential analyses of Fortunati (1995, 1998) and Katz (1999), who used semantic differentials and perceptual mapping, respectively, to understand the psychological meaning of the telephone.

A survey instrument consisting of a variety of questions related to technology and communication was constructed, pretested, and refined. It was then administered to a large class of junior-level undergraduate communication majors at a research university (most students were 20 years old, plus or minus 1 year).

Study 1 involved the readministration of this instrument (only slightly amended) to nearly 200 sophomore undergraduates at the same university. During this readministration, students were divided into two groups to compare their attitudes toward personal communication technology. The first group was asked to give opinions about personal communication technologies (PCTs)—the mobile phone, email, TV, and postcards were chosen because they vary in medium, cost, and intensity. The second group was asked to give opinions about personal *non*communication technologies (PTs)—the wristwatch, Game Boy, and Walkman were chosen because they vary in information and entertainment dimensions—compared with the mobile phone. We did this so that we could see if they were affects that might be attributable to the communication nature of the technology compared with the entertainment and personalness of the technology itself. We anticipated that by mixing the two categories together we would be less able to provide a control for the mobile phone baseline relative to the other technologies.

In Study 2, we compared the previously-described U.S. communication students with a sample of communication students at a Korean university who were also trained in English, so that the instrument did not need to be translated. (The students were located via one of the coauthors.) As an intriguing but numerically diminutive point of comparison, we also administered the English-only survey to a small number of Norwegians and Namibians in their respective countries. These data were collected by students who were traveling abroad during spring break. Because of the minuscule nonrandom samples, we present these data only to stimulate discussion and highlight the potential value of further investigation, rather than claiming scientific validity for them.

It is important to point out the limitations of the study. The purpose of the research was entirely exploratory. It is a not a study of a representative sample of communication students, university students, Americans, Koreans, or anybody else. Rather it is an analysis of two groups of students. Thus, although we do at points refer to confidence intervals and statistical significance, we do so only as what we hope will be helpful descriptors of the midpoints and spread of the data to highlight similarities and differences among variables, not as inferential measures drawn from a randomly drawn sample of a population.

3. RESULTS

Study 1: Coherence of Folk Views of the Mobile Phone.   Table 8.1 presents the adjectives that were selected from discussions in the focus group. Figure 8.1 shows how the U.S. students rated mobile phones in terms of adjective pairs. The box plots represent the mean rating and the spread of score ratings for each adjective pair. (Because of the exploratory nature of the study and for clarity of presentation, a 50 percent confidence interval was used in preference to a 95 percent confidence interval.)

Figure 8.1 shows that, at least among these students, the mobile phone may be characterized as being a technology that is popular, beautiful, and exciting. It is seen as something of a luxury and stylish, though not extremely so. As to whether it expresses the individual owner and whether it has masculine or feminine traits, the consensus seems to cluster around the midpoint; that is, there are not strong views toward either extreme.

We next used a categorical data analysis routine in SPSS (Euclidean distance model) to derive a two-dimensional stimulus configuration. Figure 8.2 shows the semantic differential items in relationship with one another on two underlying di-

TABLE 8.1
Semantic Differential Items

popular	unpopular
enemy	friend
hard	easy
boring	exciting
ugly	beautiful
jailer	liberator
cheap	expensive
masculine	feminine
work	play
secure	insecure
expresses me	does not express me
luxury	necessity
stylish	not stylish
convenient	inconvenient
useful	useless

mensions (which are statistically defined in a way that shows consistency among choices as opposed to a predefined conceptual dimension).

Several aspects of this chart are noteworthy. Luxury–Necessity appears to form its own conceptual category, consistent and unique along both dimensions. This suggests that the work by Katz and Batt (Katz, 1999) highlighting this dimension as an important and consistent latent folk category continues to be borne out with a different sample and is meaningful even when considered relative to the

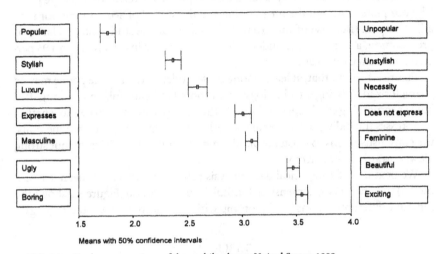

FIG. 8.1.   Student perceptions of the mobile phone: United States, 1999.

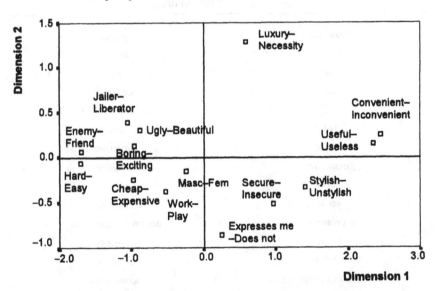

FIG. 8.2.   Saturated perceptual map of adjective pairs (Euclidean distance model).

other characteristics of the mobile phone. Convenience–Inconvenience and Usefulness–Uselessness also seem to cohere together well. Indeed, they might be seen as occupying a similar semantic space. Stylish–Unstylish and Secure–Insecure also seem to be placed near one another. This is an intriguing relationship given the body of work that has shown that teen (and adult) emphasis on fashion (style) may be related to concerns about feelings of psychological security, "belongingness," and self-esteem.

Masculine–Feminine (Masc–Fem) is near the origin of the two dimensions, suggesting that this may not be a meaningful category. Within the context of recent emphasis on "gendered technology," this finding, or, more accurately, "nonfinding," has potential implications worthy of further investigation.

Looking at Dimension 1, the adjectives Jailer–Liberator, Ugly–Beautiful, Boring–Exciting, and Cheap–Expensive all occupy the same conceptual space, though they differ somewhat on Dimension 2. A second cluster of meanings is Enemy–Friend and Hard–Easy, and both are centered on Dimension 2. Being nearly synonymous, Masculine–Feminine, Work–Play, and Cheap–Expensive fall in the same quadrant.

A second pass of the data using optimal scaling shows an interesting set of relationships among the specific answers to the semantic differential questions. Figure 8.3 shows the relationship among the specific answers to two of the semantic differential questions (Stylish and Luxury) by the status of mobile phone ownership. As may be seen in the lower left quadrant, there is a distinct cluster pattern among those who have had a mobile phone for more than a year (mobile >1 year): they tend to see the mobile phone as stylish; they also see it as a luxury and/or a necessity.

The possible conceptual U-shaped distribution of the values of longtime users may be noteworthy. There is an intriguing unresolved question—if the finding is confirmed—concerning the relationship between Luxury and Necessity. Are long-term users are of two types? The first type might be those who view it either as a necessity or as a luxury. The second type might be those who have a dual view, namely they embrace both qualities in their attitude—seeing the technology as a "luxurious necessity" (e.g., having a wristwatch but choosing a luxury brand). It is worth pointing out in this context that, although academics like to have neat conceptual categories that are mutually exclusive and exhaustive, most lay people are not so rigorous in their thinking, especially if they have not pondered and argued a question. On the other hand, it is important to emphasize that we are investigating the emotional meaning of the technology to people, rather than rigorous (or even "folk") theories.

At the opposite end of the continuum, those who have never owned a mobile phone see the device as completely not stylish. This result is intriguing and significant from several aspects, including the marketing of mobile phone technology and understanding the adoption patterns of new technology.

The data also suggest that recent adopters, relative to longtime users and non-adopters, are in the middle range of attitudes toward the stylish and luxury quali-

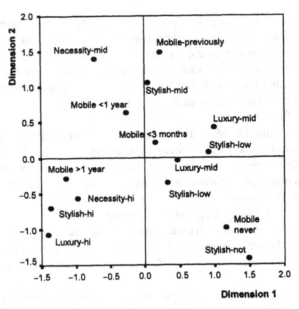

FIG. 8.3. Perceptual map of Luxury–Necessity and Stylish–Unstylish adjective pairs by
mobile phone use history (Euclidean distance model).

ties of the mobile phone. Interestingly, relatively high-status characteristics are
assigned to the mobile phone by former users and users who have had the device
for 3–12 months. It may be that nonusers who have high style evaluations of the
mobile phone were also relative longtime adopters (because they had one at some
time prior to the survey) but gave it up owing to cost considerations (college students are known for their constrained budgets). In sum, one of the strongest pre-
dictors of mobile phone ownership within this battery of attitudes is the
respondent's view of the stylishness of the technology.

We extended the semantic differential technique to explore differences among
technologies. Hence we separated out and probed the meaning of:

- Personal communication technologies (TV, email, postcard).
- Personal *non*communication technologies (Walkman, Game Boy, wristwatch).

We used five factors, two of which we report here (Expensive–Cheap; Stylish–Unstylish). The results are shown in Figs. 8.4 and 8.5, respectively.

On the Expensive–Cheap dimension, Fig. 8.4 shows that mobile phones are
seen as the most expensive of the technologies, whereas e-mail and postcards are
the cheapest. Watches and Game Boys are also considered expensive, though not
quite as much as mobile phones.

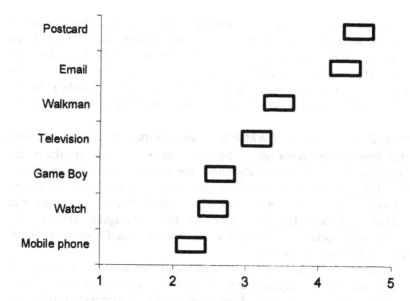

FIG. 8.4. Attitudes toward communication technologies: Expensive–Cheap dimension. *Note:* The bars represent the 95 percent confidence interval for the mean. A score of 1 represents most expensive, a score of 5 least expensive.

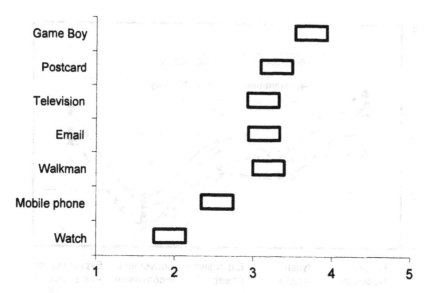

FIG. 8.5. Attitudes towards communication technologies: Stylish–Unstylish dimension. *Note:* The bars represent the 95 percent confidence interval for the mean. A score of 1 represents the highest stylishness, a score of 5 the lowest stylishness.

In terms of stylishness (Fig. 8.5), the watch is seen as by far the most stylish personal technology, and the Game Boy the least. (We speculate that the Game Boy is perceived as juvenile entertainment, from the point of view of the relatively sophisticated tastes of a young adult.) The next highest rated technology in terms of stylishness is the mobile phone, and, among the PCTs, it is rated as the most stylish. TV, email, Walkman, and postcards are all in the middling rank of stylishness.

Study 2: Perceptions of the Mobile Phone in Korea and the United States (and in Namibia and Norway).

Figure 8.6 shows that in Korea the mobile phone is seen as being about equally convenient, but more expensive and slightly more stylish, relative to the United States. It is perceived as much more of a necessity in Korea than in the United States. This might be because of the lower installed base of residential telephone services in Korea. It might also be due to the possibly greater social importance of social connection and communication to Koreans or among Korean teens.

Comparing the United States with Namibia and Norway, which is done only for speculative interest, it appears that in both the latter countries mobile phones are seen as more of a necessity and less stylish; on the other dimensions they are approximately the same.

What might be more interesting than the differences among these countries are the similarities. If it is indeed the case that young people in universities in countries as diverse as Korea, the United States, Namibia, and Norway tend to see the

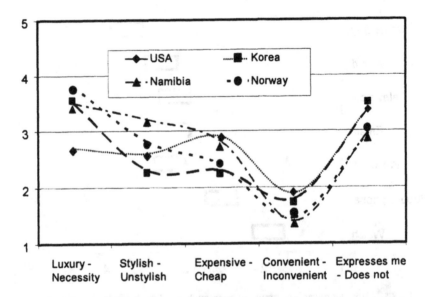

FIG. 8.6. Cross-cultural comparison of attitudes toward mobile phone characteristics. *Note:* Mean rank of adjectives by nationality (scale 1–5), where a score of 1 represents the first adjective and 5 the second adjective of the pair.

mobile phone in the same light, several explanations may account for this phenomenon, all of them intriguing. Space limitations preclude extensive discussion of them, but some of the competing and overlapping possibilities are that:

- There is an international culture of the mobile phone that spans continents.
- There is an international teen culture in which the mobile phone plays a role.
- There are universals or near-universals in the way people perceive the role of communication in their lives.

4. COMPARISON WITH OTHER RESEARCH

As mentioned earlier, we are by no means the first to use semantic differentials to understand the mobile phone. For example, in the 1990s, Fortunati had respondents compare the traditional wireline household telephone with the mobile phone. One of her samples consisted of 100 university students. They found the mobile phone dynamic but not indispensable. The fixed phone had a stronger, more articulated image. Five points stood out for the wireline phone: comfort, indispensability, speed, pleasantness, and strength.

In a second study, Fortunati (1998) investigated the attitudes of people in various age groups toward the wireline telephone, the mobile phone, and the telephone conversation. As part of this study, a semantic differential was given to 10 groups, differing in age and gender (totaling about 500 subjects). Fortunati found that men and women of different generations had different approaches to the mobile phone in terms of specific details, although their general assessments were quite similar. Of course, these studies were conducted during the second wave of the spread of mobile phones, and now that the technology is in its fourth generation it is possible that opinions have changed.

5. CONCLUSION

Some interesting possibilities are suggested by these findings. In relation to the research questions at the beginning of this chapter, it may be that there is substantial value in seeking to complement the functional and structuration perspectives with one in which folk norms and symbolic communication needs are underscored. Although space considerations prevent a more thorough examination of these implications here, we can at this point propose several conclusions:

1. There appears to be a coherent folk-meaning structure in the minds of at least one group of U.S. college students, at one university, at one time.
2. Style and fashion considerations appear to be helpful constructs in understanding user adoption of mobile phone technology.
3. There are both some remarkable consistencies and differences between cultures. Because much of academic activity in intercultural analysis and cross-cultural communication is based on differences between cultures, rather than

similarities, these findings may prove illuminating and provocative. Among the intriguing implications are that, whereas ICTs may be linked to body and fashion concepts in academic theory, they also seem to be linked in a practical and vernacular way. If this can be demonstrated, then important ramifications follow for the design and integration of ICTs into everyday life. This is a line of inquiry that the research team hopes to explore in the future.

4. Finally, we find some additional support for what we call the *Apparatgeist* ("spirit of the machine") model of ICTs and human behavior (Katz & Aakhus, 2002). That is to say, evidence suggests that people have a personal relationship with the technology, and may see it in an emotional and self-extensive way. In some ways, this may be a complement to Hegelian historicism and symbolic approaches to interpreting the built environment.

ACKNOWLEDGMENTS

We thank Leopoldina Fortunati and Carl Batt for their helpful comments.

REFERENCES AND BIBLIOGRAPHY

Aakhus, M. A. (2001, April). *Understanding ICTs in everyday life: Personal communication and information infrastructures.* Paper presented at the Machines That Become Us conference, Rutgers University, New Brunswick, NJ.

Dominoni, A. (2003). Aesthetics in microgravity. In J. E. Katz (Ed.), *Machines that become us: The social context of personal communication technology.* New Brunswick, NJ: Transaction.

Dudley, K., Steinfield, C., Kraut, R., & Katz, J. E. (1993, May). *Rethinking household telecommunications.* Paper presented at the International Communication Association Conference, Washington, DC.

Fortunati, L. (1995). *Gli italiani al telefono* [The Italians on the telephone]. Milan: Franco Angeli.

Fortunati, L. (1998). The ambiguous image of the mobile phone. In L. Haddon (Ed.), *Communications on the move: The experience of mobile telephony in the 1990s* (pp. 121–138). Stockholm, Sweden: Telia.

Frissen, V. A. J. (2000). ICTs in the rush hour of life. *The Information Society, 16,* 65–75.

Haddon, L. (2003). Domestication and mobile telephone. In J. E. Katz (Ed.), *Machines that become us: The social context of personal communication technology.* New Brunswick, NJ: Transaction.

Katz, J. E. (1999). *Connections: Studies in the social and cultural aspects of the telephone in American life.* New Brunswick, NJ: Transaction.

Katz, J. E. (Ed.). (2003). *Machines that become us: The social context of personal communication technology.* New Brunswick, NJ: Transaction.

Katz, J. E., & Aakhus, M. A. (2002). *Perpetual contact: Mobile communication, private talk, public performance.* Cambridge, England: Cambridge University Press.

Orlikowski, W. (1992). The duality of technology: Rethinking the concept of technology in organizations. *Organizations, 3*(3), 398–427.

Poole, M., & DeSanctis, G. (1992). Microlevel structuration in computer supported group decision-making. *Human Communication Research, 19,* 5–49.

Rice, R. E., & Webster, J. (2002). Adoption, diffusion, and use of new media. In C. Lin & D. Atkin (Eds.). *Communication technology and society* (pp. 141–163). Cresskill, NJ: Hampton.

CHAPTER NINE

Mobile Phone Tribes:
Youth and Social Identity

Claire Lobet-Maris

1. PREMISE

A survey carried out in October 2000 by a producer in the Global System for Mobile Communications (GSM)[1] leads us straight to a group of GSM users. It tells us
about their buying habits, their use of GSMs and short-message services (SMSs),
and what they spend. But who are these young people? This is the first question
that the sociologist asks. What social circles do they belong to; what is their school
background, their daily social and family life? A contextualization of the survey
would have been extremely useful for analyzing the data collected because, behind the sometimes staggering figures on phone calls, SMS messages, and above
all money spent on GSM, are young people whom we would like to know more
about. Behind GSM there hides a problem of social identity.

Sociologists are neither salespeople nor decision makers, but they are prudent
analysts of social phenomena on which they attempt to shed light, in order to help
both salespeople and decision-makers get their bearings before acting. It is in this
light that my chapter should be read.

2. THE MOBILE PHONE AS SYMBOLIC OBJECT
FOR THE "UNPLUGGED" GENERATION

For young people, the mobile phone is not only a practical object with which one
communicates but also an object invested with a high degree of symbolic significance and a large factor in the constitution of one's personal identity. It is undoubtedly among this age group that mobile phones have penetrated most deeply
and spread most rapidly. Network operators and telephone manufacturers have

---

[1]The survey was carried out by Motorola-Inra on a population of 300 young Belgians aged
12–18 years.

understood this, as witness the radical change in the tone of their broadcast advertising over a period of only 2 years. Just a few years ago, mobile phones were still being sold as the indispensable weapons of those warriors of modern times, the young urban professionals (yuppies). Today a new tribe is the object of advertising, the tribe of the young. A simplistic analysis might imagine that this is no more than imitative social behavior by young people who want to be like yuppies, or that it is an example of an advertising campaign that was more successful than had been expected in a way. But behind the widespread adoption of mobile phones by young people are more profound, sometimes more painful, social phenomena that we must attempt to understand.

3. MOBILE PHONES AS PERSONAL IDENTITY BUILDERS

The poll asked young people about their choice of mobile phone brands. Apart from certain considerations of price, which influence purchases among young people, it is striking that it is the style of the mobile phone that motivates young people's choice of one phone over another. Their choice is based not on functionality or even on the quality of the item purchased, but rather on the identity or image associated with particular models from different manufacturers. Associated with each style of phone is an image of youth, of sports, of rap artists, and so on, with which young people can identify and through which they begin to construct their social identity.

It is important to investigate in sociological terms this symbolic and affective investment in mobile phones by young people, to the extent that it refers us back to other, deeper questions about the gaps experienced by some young people in the assistance they receive, as they construct their social identities, from the structures surrounding them, whether these concern their schools, their families, or their social groups. Thus, quite beyond any commercial success, we must investigate the social fissures in these structures of construction that cause so many young people to attach importance to mere commercial brands, that is, to having rather than to being.

4. MOBILE PHONES AS OBJECTS OF GENDER EQUALITY

Looking back at the history of information and communication technologies (ICTs), it is important to note, in view of the results of the survey, that for the first time an ICT has spread equally among males and females. This strongly separates the mobile phone from other technological innovations—the Internet for example—where a large majority of users are still male. Two hypotheses suggest themselves in this regard. The first hypothesis is that the mobile phone is a communications technology, and therefore part of "being together," a social value that is given a high priority by females (in polling we conducted on a sample of 6,000 young people in 1994 about their relationship with computers). The second hypothesis concerns the path toward adoption of the mobile phone among young females, a path that is not the same as the one followed by males who become mobile phone users. Among young girls, the mobile phone apparently is first acquired with parental approval, as an additional security measure to guarantee the sort of protected autonomy parents desire for their daughters. The events leading to mobile phone use by young boys are different. It seems to be a personal decision to get a mobile phone, seen as a masculine tool or an object pertaining to the masculine identity, which is inferred from the way boys wear their mobile phone on their belt, like a weapon or tribal fetish. Girls do not display their mobile phones, but conceal them in special purses or holders made for that purpose.

5. TELEPHONE NETWORKS AS LOCI OF TRIBAL BELONGING

The Motorola-Inra survey does not study the reasons for choosing mobile phone networks or operators. This is worth a brief sociological excursus, for the interesting social phenomena we can find.

In choosing a mobile network, young people are motivated less by price than by image. For these young people, each mobile network has a well-known social connotation that is usually reinforced by advertising: One is humanitarian, another professional, another emphasizes families and getting together, another is provocative. Thus for youths and among youths, new groups of social belonging are rapidly created through these networks. Youths are 0475 or 0496 in the way other people belong to tribes, and accordingly there are certain rituals and codes of usage, licit and illicit, which differ from network to network.

By choosing one mobile network over another, young people are trying to construct their own social belonging to a group that shares certain values or reference points. And this social membership is reinforced by the price structures used by various network operators, which aim at "capturing" individual young people and keeping them within the network tribe by making it very costly to communicate with members of other tribes. This belonging phenomenon was closely observed by Fize (1999):

> This appears to be the response to a defective social connection, or to the lack of collective meaning of post-modern society. Adolescents recreate the ties between members of a community, making patches in a social fabric which is still tearing itself apart. Thus there appear on the scene tiny communities which operate on the principle of excluding outsiders, and these are just so many responses to a social system that does not connect anyone to anything any more, or does so with ever-decreasing effect. (p. 181)

The network becomes the place where young people arrange closed group territories for bands and groups that share the same values and the same social points of reference. Within these groups, rituals and codes circulate but they remain within the network; that is, they exist for the tribe. In order to reinforce the connection between members of such a tribe, pagers appear to be an important factor, and the survey confirmed this. First of all, pagers are cheaper than telephones and thus certain cost barriers are lowered that kept some young people from joining the network. But even more (and this was shown in the survey), pagers allow tribe members to develop secret linguistic codes, which only they know and which reinforce their feelings of belonging.

6. EXISTENCE ON THE NETWORK

One of the striking findings of the inquiry concerns the motivation of young people with regard to mobile phone usage. One might have expected the primary motivation of the young to be practical, to concern the need to communicate, to speak with others. Nothing of the sort. For a large majority of those questioned, the simple fact of being "reachable" is the most important social motivation of all. Behind this motivation can be found two important sociological phenomena that are worth studying.

The first concerns the social existence of youths within their tribe or network. In order to exist socially on a network, you have first of all to be sought out, that is, called up. Calls received are seen by young people not as functional signs, as in "I communicate a lot," but as symbolic or existential signs, as in "I am somebody in the tribe." At this level, too, pagers provide an important remedy to those who have difficulty maintaining their existence on a network. Such reasons for the rapid adoption and diffusion of pager technology were discovered by the survey. Thus, when young people are questioned about their mobile phone usage, it is impossible not to be struck by the high symbolic value they place on the number of calls they receive (whether by voice telephony or by text messaging); they are a sign of their social importance.

The second phenomenon concerns the effect on family relationships. Heurtin (1998) observed: "Family structures are changing today ... and these changes may be going in the direction indicated by the development of mobile telephone" (p. 48). Once upon a time, the apartment or the house was the spatial and social locus of communication between parents and children. Today, it is an empty area as regards social relationships. This is the result of several developments, such as the increasing number of working women and mothers, the transformation of family structures with the growing number of single-parent or patchwork families, and the increasing length of time during which parents and children live "together." All this contributes to making mobile phone and wireless messaging links more and more important as part of the relationship between parent and child. Because a place to get together no longer exists, togetherness is recomposed in a place where the other can be reached even if he or she is not there. Thus for some adolescents, the mobile phone may represent a sort of leash that constitutes a connection with parents, although the length of the leash is the object of constant negotiation: "Leave your mobile phone turned on"; "At least leave your pager on"; and so on.

It is important to stay with this phenomenon of the presence of mobile phones as part of family structures, in order to underline two aspects. First, ordinary telephones and later mobile phones and pagers became ways for young people to kill time. In just a few years the nonmobile telephone ceased to be merely a useful thing, and became a leisure-time appliance just like the television or sports equipment, filling an emptiness that neither family nor associative structures could fill. Young people communicate all the time, and are always somewhere else. The present moment is always being split, always shadowed by an "elsewhere" that takes up every moment of time not filled by some immediate activity. There is no more empty time. The "mobile" fills all the previous gaps between activities. By the same token, we have the impression that the act of transmitting and receiving messages has become as important as the messages themselves. Being connected to the network is the new way of being there, the new way of being oneself. The user is always in the process of escaping from the present, slipping away stealthily by sending a discreet little message "just between us." Studying these customs is also a way of understanding, "from the inside," the addictive behavior and psychological dependencies that can be observed with regard to mobile phones.

This question of the relation with time was particularly pointed out by Fortunati (2000), especially with regard to the impact of the cellular phone on interpersonal relationships. So, according to Fortunati (and in line with the approach of the French philosopher Bergson): "In the case of interpersonal communication managed by mobile, this expansion of time could have very harmful effects on social relations and especially love relations. In the same ways as communication feeds also on silence, so also seduction feeds on absence: if we eliminate absence and silence, we expose relations to early deterioration" (p. 4).

The other aspect to underline is the progression from nonmobile (or ordinary) to mobile phones. As Heurtin (1998) emphasized, the nonmobile telephone is a public communication tool within the family structure. Not only is it visible and audible, but its use is even traceable through telephone bills. The mobile phone, in contrast, is par excellence the tool of a certain individualization of communication that permits young people to recapture a degree of autonomy and to find a more discreet refuge from parental control in their communications. With wireless messaging a further step is taken, because it is possible by that means to communicate without speaking.

These developments should be carefully studied. Although mobile phone service providers may be celebrating those external factors that allow young people's mobile usage to fit into their parents' purchasing behavior, sociologists are getting worried. For beneath the surface of this mobile phone usage there is first and foremost a rather desperate search for social existence, for a social connection in a world that appears less and less communicative to youth. For such reasons, as the research shows, about half of young people apparently would be willing to receive wireless advertising messages. These are all indications, not likely mistaken, of the need for communication that some young people feel today. The rise in importance of mobile phones and pagers among young people is perhaps the most convincing sign of a "disconnected" society.

REFERENCES AND BIBLIOGRAPHY

Fize, M. (1999). Naissance de la culture adolescente [The birth of adolescent culture]. In *L'identité, l'individu, le groupe, la Société* (pp. 43–57). Paris: Sciences Humaines éditions.

Fortunati, L. (2000, June). *The mobile phone: New social categories and relations.* Paper presented at the International Conference "Sosiale Konsekvenser av Mobiltelefoni," Oslo, Telenor, Norway.

Heurtin, J.-P. (1998). La téléphonie mobile: Une communication itinérante ou individuelle [The mobile phone: Itinerant or individual communication]. *Réseaux, 90,* 22–43.

CHAPTER TEN

Fashion and Vulgarity in the Adoption of the Mobile Telephone Among Teens in Norway

Richard Ling

"It is 'in.' It doesn't matter if you need it or not, just that you have it."

The ownership and use of mobile telephones have become a commonplace part of the social landscape in contemporary Scandinavia. Indeed, the proper type of mobile phone used and displayed in the proper way is of importance in certain social contexts. Thus, the mobile phone is an element in the very presentation of self.

This chapter describes the adoption of mobile telephones. Specifically, it looks into their role as a type of fashion—and the ways in which they were a breach of fashion in the period 1997–1999. This period saw the rapid adoption of the mobile phone by large numbers of Norwegian teens, the penetration rate jumping from about 30 percent to 80 percent among 13- to 20-year-olds (Ling & Yttri, 2002).

These statistics are the background against which this chapter has been developed. The real question here is the adoption and use of the mobile telephone as a type of fashion statement among these teens. Those viewing the adoption of the mobile phone by teens often pose the question "What do they need it for?" This is not directly the issue when considering fashion. In this case, the statement by the father cited in the epigraph of this chapter is more to the point. It is not a question of need; rather it is a part of the individual's "personality kit" (Fortunati, 1998; Kahlert, Mühe, & Brunner, 1986; Kaiser, 1990; Landes, 1983; Tortora & Eubank, 1989).

For many teens, the ownership and display of a mobile telephone are important aspects of their lifestyle. However, like anything else, this consumption must be done correctly. As we see later, norms have developed prescribing how one should have and display a mobile telephone. Not surprisingly, the insight into these norms comes from when they are breached.

In this chapter, I first examine some of the theoretical approaches to the use of fashion and display. Following this, I return to the issue of mobile telephony and discuss it from two different points in the adoption cycle. In this analysis, I draw on the comments of teens and their parents. Finally, I place these observations into the context of style, the breaching of style, and the role this process plays in the development and maintenance of groups.

1. INDIVIDUAL INTENTION, SOCIAL INTERACTION, OR REFLEXIVE DESCRIPTION: THE DEFINITION OF FASHION

One of the first tasks of the chapter is to try to place fashion and personal display into some sort of social context. In this connection, one can posit several different aspects.

At the first level, there is the use of clothing as the expression of personal intention or status. In this modus, individuals and their choices regarding clothing are central (Cunningham & Lab, 1991; Davis, 1985; Dichter, 1985; Duncan, 1970; see also McCracken, 1988). The individual conceives of a presentation and then "passes it off" on others in more or less good faith. Cunningham and Lab implicitly included this when they noted: "Clothing helps to define our identity by supplying cues and symbols that assist us in categorizing within the culture" (p. 11). It is perhaps appropriate to draw on Goffman, who has pointed to this in his analysis of front and back region performances (Goffman, 1959, 1967).

In many respects, we are now leaving the intention of the individual and entering into a more complex discussion of fashion at the next level. It is here that we find the interaction between the person's presentation of his or her self through the use of dress and the observer's interpretation of the individual (Forsythe, Drake, & Hogan, 1985; Harp, Stretch, & Harp, 1985; Rucker, McGee, Hopkins, Harrison, & Utts, 1985).

At the next level, there is the notion of the interaction between the individual and the "viewer" of the clothing or artifact. Thus, one introduces the idea that the wearer of the clothing, in effect, makes a statement that is then interpreted by the viewer. The interpretation may be true to the intention of the wearer, or perhaps it is clouded in the bias of the viewer. None the less, clothing and fashion in this perspective are seen as an interaction between the wearer and the viewer. Just as with conversation, there is an exchange of communications between the individuals, and there is a whole range of para-communication or background communication that goes on in parallel with the more open interaction (Ling, 1998).

Planting ourselves firmly in the social and the interactive dimensions, the person who has carried out a key analysis of fashion is Georg Simmel.[1] According to Simmel, fashion has several dimensions that mix to give the striving for (or after) fashionability a dynamic nature. In fashion, one finds the blending of desire for individual statement and, at the same time, the seemingly opposite desire for group identification: "Two social tendencies are essential to the establishment of fash-

---

[1]Simmel has been criticized as being too class based in his analysis (Davis, 1992; Polhemus, 1994; Sproles, 1985). None the less his analysis is interesting for being dynamic.

ion, namely, the need of union on the one hand and the need of isolation on the other" (Simmel, 1971, p. 301).

Among teens, there is the search for identity in collaboration with their peers and the need to mark differences from others, in particular from other generations but also from other "types" of teens (Lynne, 2000). Within adolescent culture, and in fact within culture at large, there is the need to mark the boundaries between groups (Flugel, 1950). Clothing and other types of artifacts are used in this activity (Douglas & Isherwood, 1996). Through the adoption of the fashion of the group one becomes an example of that group and, in this way, becomes the embodiment of "a joint spirit" (Simmel, 1971, pp. 304–305).

Beyond simply marking membership, such displays can be the basis for both inclusion and exclusion. Fashion can be used graphically to display one's allegiance to a group, or it can mark one out as a "legitimate" target for retribution.

Looking for a moment at the opposite case, exclusion can appear when considering the temporal and economic aspects of personal display. Specifically, those who strive after the adoption of a style or a fashionable artifact such as a mobile telephone and who arrive "late at the party" as it were may never actually fulfill the full potential of the display. They may be categorized as "wannabes" or nouveau riche, and they can never share in the social bond as described by Duncan (1970). By the time that those who strive after fashion have arrived, the cognoscenti at the cutting edge have likely moved on to something else. Fashion is thus a balancing of the future and the past. If one is too far in front or too far behind popular taste then one is out of fashion. Indeed, Simmel (1971) suggested that fashion is the dividing line between the past and the future—that which was and that which is coming.

A final approach, which is more thoroughly examined in the conclusion of the chapter, is the interpretation of another's clothing and artifact display within the viewer's peer group. It is important to emphasize that this is not an interaction between the person displaying the clothing and a viewer or with the person and their choice of clothing. Rather, it is the use of that person's presentation of self to enhance the internal cohesion of another group.

2. THE UNDERSTANDING OF THE MOBILE TELEPHONE AMONG TEENS

Turning now to the empirical material, the mobile telephone is used to illustrate some of the issues outlined previously and as a springboard towards the examination of peer group integration.

In the late 1990s, the price of mobile telephony was falling, particularly the up-front cost of the handsets. This was followed in the fall and winter of 1997/ 1998 by the introduction of the prepaid subscription. These two elements combined to allow teens access to the system on a popular basis.

We conducted two sets of interviews. The first set of material comes from the summer of 1997, the period when mobile phone handsets were subsidized but immediately before the commercialization of prepaid calling cards. Thus, mobile telephones had started to appear on the scene, but they were not by any means widespread. The second set of material comes from the period after the widespread adoption of mobile telephones by teens in 1999 and 2000. Rather than being a novelty during this latter period, the device was an everyday part of teens' lives.

Before Widespread Adoption.   In the summer of 1997, ownership of a mobile telephone was still exceptional and new, as seen in the following sequence:

Interviewer: OK. Can you remember, is this something that has happened in the last year?
Bente: It is now, in the last month of the school year because it has been so "in."
Interviewer: OK. In the last month.
Bente: It is now that mobile phones are "in" you know. So all the time since Christmas it was pagers but now it is mobile telephones.

Teens' ownership of the technology was still unique and had not completely been domesticated. This is seen in Bente's comment that her friends took their mobile telephones with them to school more for effect than for any functional purpose:

Bente: Now it is "in" to have a mobile telephone. A lot of my friends have gotten a mobile telephone. So, they are at school just to show them off.

Bente reinforces her estimation that the device is new by repeating that it is "in." Her perhaps ironically intoned notion that it is "in" indicates that she is keeping a critical distance from the trend. In her comments she brings up the point that her friends had them at school "just to show them off." Her father, who was also present during the interview, adopted somewhat the same line: "It is in. It doesn't matter if you need it or not, just that you have it." These comments point to a heightened sense of the moment, as with Simmel's talk of the tension between the avant garde and mass adoption.

At this early point in the adoption of the mobile telephone among teens, public use of the device did not go unnoticed. Its strong connection with the "yuppie" culture and its expense marked its public use or display by a teen as a special event worthy of comment. One can see this in the comments of Martin, a 17-year-old respondent:

Martin: ... There are a lot of them that use it, like if you go places where there are a lot of people like, and like you can go "Yeah, hi." Shout loud like. [They] can almost hold their hand up to the mobile telephone and point like "I have a mobile telephone." That is, like, shouting to show that he has a mobile telephone.

There is the sense that the mobile phone is being used in a vulgar, unsophisticated way, to attain status. Given this general orientation, one can note how both Bente and Martin characterize the use of the device. As we see in the discussion, both had a codex of terms available (Fine, 1987):

> Bente (13): I don't want one because I think that it is a bit tacky [*harry*] to have one. There are a lot who think it is cool. So I don't have one.

Bente employs the Norwegian term *harry*, which means "tacky" or "vulgar." It is the opposite of "cool." Vulgarity is often associated with the sense of display that is above one's station. Thus, one is asserting a "face" in the Goffmanian sense, but the display is only partially successful.

Martin too is able to place a characterization on the display of the mobile telephone:

> Martin (17): There are some in my class you know, there are a lot that are pompous (*soss*) and [wear] really expensive brand-name clothes and things, and many of them have, you know, a mobile telephone.

Instead of *harry* Martin uses the Norwegian term *soss*, which means "preppy" or "pompous" (Lynne, 2000). It contains some of the same elements as *harry* from a slightly different perspective. Both are derogatory adjectives that are used to describe those who appear to seek status through personal display. There is also often a class-based element in the use of *soss* and *harry*. *Soss* often refers to those who are perceived to be above the observer's station, whereas *harry* often refers to those who are below.

As with many derogatory terms, these are quite loose and flexible in their application. Indeed their looseness and flexibility are essential to their role in the vocabulary. Their power lies in the fact that different people can stretch them to cover different situations. This very flexibility makes them dynamic words whose application itself is a topic of interest.

After the Widespread Adoption of the Mobile Telephone Among Teens.

In the interval between the first set of interviews and those carried out in 1999/2000, teens' use of the mobile telephone had become a more common sight. Beyond the mere ownership statistics, the device attained a currency among teens representing both access to friends and an aspect of the child's emancipation (Ling & Yttri, 2002).

Once the threshold of whether or not to be seen with a mobile telephone was crossed, the type, price, and size of the mobile phone were often seen as important. Whereas in 1997 there was a broad notion regarding style, by 1999/2000 this had became precise.

Comments from parents of the teens indicate that the style and type of mobile telephone have significance:

Marta: Now it is like our children will get our old telephones.
Gerd: No, they are not good enough.

Another mother reported that her daughter "wouldn't touch my [mobile telephone] in public." The daughter had to take the SIM card from her mother's handset and put it into her own before she was willing to use her mother's subscription. These comments point to the importance of fashion in relation to mobile telephony. They also point to a concern regarding the type and style of mobile telephone with which one is seen.

Again, Simmel's discussion regarding the currency of style is relevant here, along with the issues of identity and group membership. The generational difference almost guarantees that at least the nuances of how the symbols are displayed are dissimilar and, more likely, the whole codex of symbols is different.

When looking for the elements that are particularly sought after, the age and size of the mobile phone, the model, the fascia, particular colors, the functions of the handset, and sometimes various accessories such as pouches were common points of comment in many of the interviews in the second round. What is important is not simply owning or not owning a device, but rather its appearance.

Another characteristic of the second time period is that the teens had developed notions of how one should display a mobile telephone. The watchword here seems to be discretion. Informants noted that "it is a bit *harry* to carry it on your belt. It is not very cool to go around showing off your mobile telephone." This quote describes their relationship to displaying the device. The adjective *harry* reappears here to describe people who carry a mobile phone on their belt. This is seen as being too conspicuous. The best strategy according to these teens is to have the telephone in your pocket or in your bag.

The informants drew on some of the same concepts when discussing how the device is used. For example, Rita (18) noted, "The most *harry* of all is when people use their telephone in the bus and talk really loud. That is just such a pain." This comment is not a general condemnation of those who use a mobile telephone Rather, the mobile telephone is accepted but the specific situation is in question.

3. *HARRY* AND *SOSS*: IDENTITY, GROUP MEMBERSHIP, AND TERMS OF DEGRADATION

As we have seen, the sight of someone using, or even owning, a mobile telephone can be characterized in a range of ways. Now we turn to the question of what all of this means. What is the effect of putting names, either degrading or respectful, on the sight of various social phenomena?

We have at our disposal a whole set of terms with which to interpret the display of others. These interpretations are dynamic, as seen in the data in the previous section. The things that constitute *harry* and *soss* change with time. At one point in time and in one location, *soss* can define someone who wears a certain type of denim jacket and style of pants. In another time or place, the same word can be applied to someone who wears their collar turned up, or wears Lacoste shirts and Treitorn shoes, and so on.

In the context of this analysis comparing two time periods, the concept of *harry* in relation to mobile phones seems to have shifted from being a general description to describing specific situations. The informants in the first round of interviews were not so able to describe the specific things that made for a stylish mobile phone. During the second set of interviews, the informants were generally mobile phone owners—as many as three fourths of all teens owned one at this point. There were fewer examples of belittling terms in these interviews and their use was more specific. The informants' perception of what was stylish was also more detailed, namely the size, color, fascia, and model of the most sought after mobile telephones. Furthermore, there had developed a style of display and an understanding of where and how one should use the device. These considerations indicate that the device had become embedded in the youth culture.

An interesting point here is the development and use of the terms to describe mobile telephone use. The notions of *harry* and *soss* have been described. Particularly in the case of *harry* they are degrading terms that are antonyms for what is fashionable, stylish, and cool. From a Simmelian perspective, rather than engendering inclusion and a sense of belonging, these characterizations point more toward a sense of exclusion. Individuals who characterize others as being either *soss* or *harry* are not including them in their social group. Indeed, these terms are not often used to a person's face because they are a type of insult.

The next point, which is somewhat curious, is that one only rarely comments on the untoward behavior of others. Goffman (1963) described this as "civil inattention." He suggested that we, as social actors, strive as much as possible to maintain the sense of the situation. Thus we try to ignore others who threaten this sense. Indeed, this emphasis on maintaining the inertia of a social setting means that one's behavior can reach the point of being false and studied. As noted by Goffman, "to behave properly and to have the right to civil inattention are related: Propriety on the individual's part tends to ensure his being accorded civil inattention" (p. 87).

The point at which intrusive conduct breaches the sense of propriety is flexible and situational. As shown in the material presented previously, it has to do with one's experience of the intrusion in question, the context in which it takes place, and the degree to which the person doing the breaching is able to alter the definition of the situation to their own ends. The implication of describing others as *harry* or *soss* is that they have overstretched the sense of proper conduct in a public place.

Up to this point, the discussion has been about how individuals relate to a situation as it unfolds, there and then. The ex post facto description of others in derogatory terms indicates that the rule of civil inattention was perhaps extended in the specific situation, but that the tension was so great as to imprint that situation on the mind of the speaker for later comment and characterization. It is this process of applying words to a situation, that is, packing it into the vocabulary of the individual and transporting it to other social situations for retelling, that is of interest here.

The specific point of departure is seeing something as *harry* or *soss*. This refers not so much to people presenting themselves as being a certain way as to others' sense of who those people are and how they present themselves. Within the specific situation, the individual is afforded the deference of not being told that they are, in Goffmanian terms, "out of face."

The process of recounting a socially awkward situation to others involves what one might call the institutionalization of characterizations. The observer interprets the presentation and formulates a description using a repertoire of words and concepts that have been built up in the individual's interaction with others. This is a critical aspect of the process. The various terms are invested with meanings depending on the experiences of the individual and the experiences of the groups in which the individual participates.

Another consideration is whether or not the third party observed the original presentation. If two people who share a general repertoire of descriptive terms view the presentation and if the two are in agreement about the proper description of the event, then the incident becomes a further confirmation of the concept's content and also a celebration of the group's identity. If the event is related, secondhand, to others who have a common perspective, the audience has to draw on its own understanding of the speaker's description of, for example, the *soss* or "oaf." This mapping of new experiences on an already existing framework of friendship is a further development of the common ideology. Thus, the assertion that a viewed person is "neat," "dowdy," "stylish," or "gaudy"—or a whole set of other characterizations—helps maintain the group ethic. The group is provided with common reference points and a common sense of how characterizations should be applied.

The use of derogatory characterizations can also represent the exercise of power. Asserting one's definition of, for example, *soss* within the context of the group is part of the process of asserting a certain power over how group members should present themselves. Such terms can take on the character of casual and perhaps humorous chatter among friends, or they can be used as serious expressions of control. The latter situation obtains perhaps more often among teens, for whom the building up and establishing of identity is a central issue. Indeed, one of the crucial issues of adolescence is emancipation from one's parents and the establishment of one's independence.

Sullivan (1953) and those who have followed his analysis have noted that the adolescent peer group is a necessary area for the development of the socialized adult. The peer group provides adolescents with a relatively bounded situation where, none the less, they can exercise certain forms of control and participate in group decision making among equals (Giordano, 1995; Ling & Helmersen, 1999; Savin-Williams & Berndt, 1990; Youniss, 1980; Youniss & Smollar, 1985). The peer group provides self-esteem, reciprocal self-disclosure, emotional support, advice, and information. This is seen in the growth of an idioculture of nicknames, jokes, styles, music, artifacts, and so on (Fine, 1987).

One of the activities of the peer group is the informal establishment of codes of dress consumption patterns and their orientation (Hogan, 1985). It may well be that there is a greater need for identification than for distinction within the adolescent peer group. In addition, the established patterns can be either mildly or more ruthlessly enforced. The preceding material offered some relatively benign linguistic examples of how style of presentation vis-à-vis the mobile telephone is enforced. The use of terms such as *harry* and *soss* can be viewed in this context.

The need to resort to derogatory comments about the behavior of others indicates that they do not share the sense of the interaction and that they are not "in line," nor are they worthy of envy in the sense that Simmel talked about fashion. There is a consciousness of the need not to be like they are. Thus, the notions of *harry* and *soss* contain ideas that are the polar opposites of "fashionable" in the Simmelian sense of the word. Rather than asserting identity and the desire for group membership, they point toward the sense of wanting to distance oneself from a phenomenon.

There is a similarity in the concepts, however, in that the specific definition of what is fashionable and what is *harry* or *soss* is constantly being redefined. Just as fashion is dynamic, so is the need to distance oneself from that which is tacky or overbearing. Whereas gold necklaces and other such disco attire were "in" during the 1980s, they are definitely "out" in the new millennium. The popularity of the mobile telephone will share the same fate.

REFERENCES AND BIBLIOGRAPHY

Cunningham, P. A., & Lab, S. V. (1991). Understanding dress and popular culture. In P. A. Cunningham & S. V. Lab, *Dress and popular culture* (pp. 5–20). Bowling Green, OH: Bowling Green State University Popular Press.

Davis, F. (1985). Clothing and fashion as communication. In M. R. Solomon (Ed.), *The psychology of fashion* (pp. 15–27). Lexington, MA: Heath.

Davis, F. (1992). *Fashion, culture and identity.* Chicago: University of Chicago Press.

Dichter, E. (1985). Why we dress the way we do. In M. R. Solomon (Ed.), *The psychology of fashion* (pp. 29–37). Lexington, MA: Heath.

Douglas, M., & Isherwood, B. (1996). *The world of goods: Towards an anthropology of consumption.* London: Routledge.

Duncan, H. D. (1970). *Communication and the social order.* London: Oxford University Press.

Fine, G. (1987). *With the boys: Little League baseball and pre-adolescent culture.* Chicago: University of Chicago Press.

Flugel, J. C. (1950). *The psychology of clothes.* London: Hogarth Press.

Forsythe, S., Drake, M. F., & Hogan, J. H. (1985). Influence of clothing attributes on perception of personal characteristics. In M. R. Solomon (Ed.), *The psychology of fashion* (pp. 268–277). Lexington, MA: Heath.

Fortunati, L. (1998). Revêtir des technologies [Wearing technology]. *Réseaux, 90,* 85–91.

Giordano, P. C. (1995). The wider circle of friends. *American Journal of Sociology, 101*(3), 661–697.

Goffman, E. (1959). *The presentation of self in everyday life.* Garden City, NJ: Doubleday.

Goffman, E. (1963). *Behavior in public places: Note on the social organization of gatherings.* New York: The Free Press.

Goffman, E. (1967). *Ritual interaction: Essays on face-to-face behavior.* New York: Pantheon.

Harp, S. S., Stretch, S. M., & Harp, D. A. (1985). The influences of apparel in responses to television news anchorwomen. In M. R. Solomon (Ed.), *The psychology of fashion* (pp. 279–289). Lexington, MA: Heath.

Hogan, D. P. (1985). Parental influences on the timing of early life transitions. *Current Perspectives on Aging and Life Cycle, 1,* 1–59.

Kahlert, H., Mühe, R., & Brunner, G. L. (1986). *Wristwatches: History of a century's development.* West Chester, PA: Schiffer.

Kaiser, S. B. (1990). *The social psychology of clothing.* New York: Fairchild.

Landes, D. S. (1983). *The revolution in time: Clocks and the making of the modern world.* Cambridge, MA: Belknap Press.

Ling, R. (1998). *"She calls, [but] it's for both of us you know": The use of traditional and mobile telephony for social networking and among Norwegian parents.* Kjeller, Telenor FoU, FoU R 33 / 1998, Norway.

Ling, R., & Helmersen, P. (1999). *It must be necessary, it has to cover a need: The adoption of mobile telephony among pre-adolescents and adolescents.* Kjeller, Telenor FoU, FoU R 9 / 2000, Norway.

Ling, R., & Yttri, B. (2002). "Nobody sits at home and waits for the telephone to ring": Micro and hyper-coordination through the use of the mobile telephone. In J. Katz & M. Aakhus (Eds.), *Perpetual contact: Mobile communication, private talk, and public performance* (pp. 139–168). Cambridge, England: Cambridge University Press.

Lynne, A. (2000). *Nyansens makt (en studie av ungdom, identitet og klær)* [The power of nuance: A study of youth identity and clothes]. Lysaker, Norway: Statens institutt for forbruksforskning [The National Institute for Consumer Research (SIFO)], Rapport 4.

McCracken, G. (1988). *Culture and consumptions: New approaches to the symbolic character of consumer goods and activities.* Bloomington: Indiana University Press.

Polhemus, T. (1994). *Street style: From sidewalk to catwalk.* London: Thames & Hudson.

Rucker, M., McGee, K., Hopkins, M., Harrison, A., & Utts, J. (1985). Effects of similarity and consistency of style of dress on impression formation. In M. R. Solomon (Ed.), *The psychology of fashion* (pp. 310–319). Lexington, MA: Heath.

Savin-Williams, R. C., & Berndt, T. J. (1990). Friendship and peer relations. In S. S. Feldman & G. R. Elliott (Eds.), *At the threshold: The developing adolescent* (pp. 277–307). Cambridge, MA: Harvard University Press.

Simmel, G. (1971). *On individuality and social forms* (D. N. Levine, Ed.). Chicago: University of Chicago Press.

Sproles, G. P. (1985). Behavioral science theories of fashion. In M. R. Solomon (Ed.), *The psychology of fashion* (pp. 55–70). Lexington, MA: Heath.

Sullivan, H. S. (1953). *The interpersonal theory of psychiatry.* New York: Norton.

Tortora, P., & Eubank, K. (1989). *A survey of historic costume.* New York: Fairchild.

Youniss, J. (1980). *Parents and peers in social development: A Piaget–Sullivan perspective.* Chicago: University of Chicago Press.

Youniss, J., & Smollar, J. (1985). *Adolescent relations with mothers, fathers and friends.* Chicago: University of Chicago Press.

CHAPTER ELEVEN

Extension of the Hand: Children's and Teenagers' Relationship With the Mobile Phone in Finland

Virpi Oksman and Pirjo Rautiainen

Young people in Finland have been particularly quick to adopt the mobile phone into their lives: 77 percent of Finns aged 15–19 have the use of a mobile phone (Nurmela, Heinonen, Ollila, & Virtanen, 2000). The phenomenon is by no means exclusively urban: Regional differences in the distribution of mobile phones are relatively small.

Today, mobile phones are commonly acquired for children aged 10 to 12. According to City of Helsinki Urban Facts, one third of 10-year-olds in the Helsinki area own a personal mobile phone. The relationship between children and mobile communication generates a variety of questions concerning issues such as the suitable age to acquire a mobile handset, child rearing, use of the mobile phone in schools, and children's relationship to the device, to mention but a few. A cultural negotiation participated in by the media, families, and professional educators is currently taking place regarding the phenomenon.

Since 1997 the Information Society Research Centre at the University of Tampere has been studying young Finns' use of mobile telephony in cooperation with mobile operator Sonera Mobile Operations and mobile phone manufacturer Nokia Mobile Phones. The National Technology Agency of Finland (Tekes) has been financing the project "Mobile Communication Culture of Children and Teenagers in Finland" since 1998.

The purpose of this chapter is to consider children's and teenagers' use of mobile telephony as a cultural and social phenomenon and to pose questions about what kind of meanings teenagers and families with children attribute to mobile communication and to the device and about what kind of communication they see as important. Young people in Finland have an unreserved everyday attitude to the mobile phone. For example, rather than referring to their handsets simply as "mobile phones," they use the words *känny* or *kännykkä*, which can be interpreted to mean "an extension of the hand."

1. THE MEDIA ETHNOGRAPHIC APPROACH

Through use of a methodology often termed media ethnography, we have aimed to bring out the meanings attributed to mobile communication by children and teenagers. A number of British scholars have used media ethnography to examine different media and communication devices in a sociocultural context, often with a focus on domestic media environments (see Morley, 1986; Silverstone & Hirsch, 1999). We have positioned the subject of our media ethnographic inquiry, the mobile phone, in both domestic and public places. One of the distinctive characteristics of the mobile telephone is that it intertwines the domestic and public spheres, for instance by bringing personal life to the office and vice versa.

The main emphasis of the research is on hearing what the children and young people themselves have to say. Since 1997 nearly 1,000 interviews have been conducted by the research group. The interviews took place in both urban and rural areas. The family background of the informants is socioeconomically heterogeneous. Aside from the interviews, various types of qualitative material were collected: observation at youth events, photographs, text message material, media diaries, evaluation of mobile portals, and children's drawings of their dream mobiles. Since 1998 the research group have gathered a collection of some 7,500 text messages.

2. A "SAFETY LINE" FOR THE FAMILY

As in other Nordic countries, the participation of Finnish women in working life has been very high. Moreover, according to Statistics Finland (2001), the number of stepfamilies and one-parent families in relation to nuclear families has increased. New domestic situations present families with new needs connected with the organizing of everyday life; the mobile phone is, in part, an answer to the demands generated by the emergence of new cultures in both work and family life.

Parents frequently consider purchasing a mobile phone for a child when the child's living environment starts to broaden outside the home. This tends to occur when the child begins school and hobbies and friends become more significant. Parents interested in new media and information technologies are more likely to purchase a mobile phone for a child, and, having bought a mobile phone for their child, the parents may perceive themselves as pioneers in the use of new technology.

Teenagers' and parents' motivations for mobile phone purchase frequently differ. As parents see it, a young person's mobile handset is acquired for security reasons or in order to improve the teen's accessibility to parents. Teens may use the accessibility argument themselves when negotiating the purchase with parents, although the actual reason for wanting a mobile handset is often the desire to keep in touch with friends.

With children, however, the purchase is in most cases initiated by the parents, and the child usually does not participate in the purchase decision. Children frequently describe the mobile purchase as an event with an element of surprise in it. A 9-year-old girl related the event of getting her first mobile phone: "Dad just came and said, here's a mobile for you." The mobile communication between parents and children is not significant in quantity; in fact, the number of calls and messages between children and their parents is rather small. More than communication itself, the parents value the possibility of contacting their children any time they wish. As the mother of an 8-year-old girl put it: "Children having the mobile with them at all times reassures me somehow. It makes you feel secure when you think that if they have a problem they can reach you anytime, anyplace."

3. A TOY OR A UTILITY ARTICLE?

The more widely the phenomenon has spread among Finnish teenagers, the more varied have attitudes to the device become and the more differentiatedly and specifically have the different user profiles evolved. When in 1997 the mobile phone began its proliferation among Finnish teenagers, the mobile device was still regarded as a status symbol. The make, the model, and the price of the handset mattered. At that time, the mobile phone also constituted a popular subject of discussion. For instance when asked in 1997, Have you talked about mobiles in school?, a 15-year-old boy replied, "Among boys in general, it's probably the most talked about subject at the moment, the mobile."

In 1997, teenagers' attitudes toward their mobile phones were respectful and cautious. They were careful to use their mobiles in the "correct" way. The object still retained the element of status: The mobile handset was a valued item in itself, enhancing the individuality of its owner without further personalization. Little by little teens began to use their mobiles like other consumer goods.

At the end of the first wave of proliferation between 1997 and 1998, parents began to acquire mobiles for children under the age of 12. The statements of the interviewees reflected appreciation of the mobile terminal: Parents found it necessary to clarify to their children that the mobile handset was not a toy but a real telephone, an appliance that should be handled with responsibility. As Anna, a 5-year-old who has had a mobile since 1998, said in an interview held in fall 1999: "First I thought it was a toy, and then my mom told me it's not something to play with, it's a phone. I didn't understand that at first, and I started to play with it. I imagined that it was a shark. That this one person came into the water, and it was a shark."

Particularly during the early years of the research, teenagers made frequent comparisons between mobile phones and virtual pets, tamagotchis. This can be seen as one example of young people's perception of the mobile phone as a living thing, which, in turn, is linked to the position of the mobile phone as a symbol representing teens' friends who can be contacted through the device. Personifying the mobile phone was evident both in rhetoric and in emotional reactions directed to the phone. A 16-year-old girl put it this way: "The antenna of my mobile broke and I freaked out completely. I kept thinking: 'Mobile, please forgive me!'"

In 2000 the mobile phone was no longer a new object in the everyday life of children; children had gotten used to the phones after seeing their parents and other people use them. Stressing the fact that mobile phones are not to be played with was no longer considered necessary in families where parents had purchased mobile phones for children.

4. HANDMADE TECHNOLOGY AS A PHENOMENON

User-replaceable color covers and other accessories and services for the purpose of personalizing the device were already available in 1997, but the young had thought up a way to challenge the industry—some of the teenagers began to decorate the phones themselves. One of the first mentions of this is by a boy in 1997, who painted his mobile to match the color of his moped. The 16-year-old boy reported: "I paint them myself, 'cause they're so expensive to buy." The same phenomenon emerged among Finnish girls in spring 1999, only more strongly: They began decorating their mobiles by painting them with nail polish. For example, a 15-year-old girl reported in 1999: "Last night I gave it [the mobile] a new coat when I was varnishing my nails."

The mobile phone and its accessories were thus shaped in various ways in order to make them part of the owner's personal style. This can be seen as an attempt to humanize and seize control of technology. Personalizing signifies breaking earlier conventions with regard to technology; high technology should not necessarily look like high technology.

In winter 1999, services connected with personalizing the device started to become more common. First it became possible to order logos for the handset screen, and then different ringing tones became available. As the phone models owned by teenagers got more recent, teens became large-scale consumers of services designed to personalize the phone. The appearance of the mobile phone served to express belonging to a certain group or to accentuate the individual style of the user.

Another change in the teens' relationship with their mobiles became evident in 2000. The young no longer spoke of the device itself. Instead, their interest focused on logos, ringing tones, and the contents of text messages. Web talk has increased significantly and to an extent replaced mobile talk. Conversations revolve around topics such as: Where on the Net did you find the coolest logo or ringing tone? What's your home page like? Today, talking about mobiles has moved from comparing the different models to comparing personalized services.

5. GENDERED TECHNOLOGY

For young people, mastering the new communication technology has become a norm. A 17-year-old boy described the importance of remaining up to date with technology: "You have to keep up really, if you want to make it in this world." Attitudes towards technology among teenagers remain gendered, however (see Brosnan & Davidson, 1994; Chiaro & Fortunati, 1998; Fortunati, 1995; Håpes & Rasmussen, 2000a; Oksman, 1999). Technology optimism is more characteristic of boys. A 15-year-old boy commented: "It [technological development] has been quite amazing in Finland in the 1990s, and it's good that it's been fast." Though actual technology phobia is rare, girls remain more reserved about developments in technology and the devices produced by communication technology. Yet most teens with reservations about technology accept mobile phones and use them, because they view mobile phones as controllable technology. Girls' fears are directed more toward computers than mobile telephony; in addition, the more extreme technological developments often intimidate girls.

Though boys and girls have been equally active in adopting the mobile phone, the material does suggest some gender differences in its use and that traditional gender roles exist in mobile culture. Boys are more interested in keeping up to date with innovations and developments in mobile technology. They frequently view the mobile phone as a tool for regulating and controlling one's environment. Girls are more focused on the interactive and aesthetic side of communication technology and in contact with geographically distant people (see Håpes & Rasmussen, 2000b).

6. PARENTS' USE OPTIMISM

In industrialized Western countries, a cultural view currently prevails according to which children are extremely interested in new technologies and become skilled users in a short space of time. Parents frequently exaggerate the ease with which children learn to use computers and mobile phones, often mentioning how their 8-year-old is better at using the phone than they are. Parents tend to mistake manual dexterity in mobile games and general "tinkering" with the phone as actual communication skills. Leopoldina Fortunati made similar observations when studying the use of the fixed-line phone in Italian families. Parents had not always explained the use of the telephone to children. Small children do, however, generally need help in learning to use the telephone (Fortunati, 1995).

According to this cultural view, new technologies will be used by ever younger age groups. In particular, the age at which the mobile phone is acquired is expected to fall. In 1997, young people generally received their mobiles as a confirmation present at the age of 15. In 2000, parents anticipated that the age of acquisition would go down to the school-starting age of 7. Parents also assumed that special reasons would no longer be necessary in order to purchase a mobile phone for a child, because they expected the culture to change so that children's mobile phones would come to be seen as casual everyday appliances: "In a few years nearly every child starting school at 7 will have a mobile phone. Going lower than that makes no sense to me, since I, at least, am unable to think of any reason

why a 3 year old should have a mobile phone when he's playing in the sandbox."
(Father of a 9-year-old girl).

After acquiring a mobile phone for a child, parents expect rational communication behavior from the child; it is a matter of honor to the parents. Parents perceive smart use of the mobile phone as prudent and generating little cost. A typical description of mobile phone use by a child characterizes the communication as economical and functional: "Niko is a very smart mobile user. He doesn't chatter. He says what he called to say and then it's bye-bye." (Mother of an 8-year-old boy).

7. CHILDREN'S RELATIONSHIP WITH THE MOBILE PHONE

Children none the less report occasional problems even with functions that they are most familiar with. Contrary to the views of their parents, children also talk about spending a lot of time learning the functions of the phone. As an 11-year-old girl put it, "I still don't know how to use everything in it. But I've tried. I browsed it a lot, and it took me a couple of years to learn to send text messages and things like that. I just tried things out."

It is not rare for children to perceive mobile games as the most interesting feature of the handset. From the child's perspective, the mobile phone may be primarily a games machine, and the communicative features (calls and text messages) may be secondary. The relationship between small children and mobile communication is often rather distant, and small children are rarely active subjects in mobile communication in the way teenagers are. Because of their use optimism, parents may be surprised about the child's mobile communication being largely different from that of adults. As the mother of a 10-year-old boy remarked, "He can call about the most curious things, on the spur of the moment, things that he finds important—like finding his keys."

Children's communication is tied to the moment and requires immediate feedback. The nonverbal side is also more emphasized than with adults: besides facial expressions and gestures, children use their whole body to deliver the message (see Wood, 1976).

A 5- to10-year-old child's relationship with the mobile phone may be practical, personifying, or imaginative.

A Practical Attitude. For children with a pragmatic attitude, the mobile phone is useful for them to be able to contact their parents, and vice versa. They do not have special nicknames for the mobile phone, but refer to it as *känny* or *kännykkä*. The device itself should, in the child's opinion, have the appearance of a telephone; a mobile phone resembling a teddy bear, for instance, would not be appropriate. As an 8-year-old boy said, when asked What would you think if the mobile was a furry teddy bear?, "I wouldn't like it, it would feel stupid, a furry mobile [contemptuously]. It should look like a phone."

Children with a pragmatic attitude do not personalize or decorate their mobile phone, and any extra covers, logos, and ringing tones will have been bought or ordered by the parents. Any enthusiasm they may have had for their mobile phone has died down quickly. In their everyday life, the mobile phone does not hold the charm of novelty; it has existed for as long as they can remember. It is not uncommon for such children to forget to take the phone along when visiting a friend, for example.

A Personifying Attitude. Children who personify their mobile phone may attribute a variety of different meanings to the device through pet names such as "Heart," "Cherry," or "Garfield." They speak of the mobile as if it were a living being with a body and vital functions. This kind of attitude is referred to by the concept of animism in child psychology (see Hurlock, 1978; Piaget, 1977). For example, such children may describe their father's technologically sophisticated handset as "real skinny, a lot more skinny than mine," or may perceive the extended battery as "a bump on its back." They may also exhibit nurturing behavior toward the mobile, dress it in Barbie clothes, or attribute qualities of living beings to it, as in the following dialogue:

This mobile has terrible hiccups.
What did you say? Hiccups?
Bad reception. I call it the hiccups. (Girl, 11)

In cases like this, the child is more interested in mobile communication and mobile games than children on average. The child sees the functions of the mobile phone as easy and picks them up quickly.

An Imaginative Attitude. An 8-year-old girl offered the following description: "I call it [the mobile] a ringing square sometimes. My friends said to me once: 'Your mobile doesn't work.' I just told them it doesn't matter, it's just a ringing square." Children with an imaginative attitude may perceive the mobile phone through their imaginary world, but the notion of mobile telephony may remain distant and the meaning of the device obscure. The child does not decorate or personalize the phone, because the device itself is too distant or strange to arouse interest or to create an emotional bond. The child has the phone on when told, but does not understand why it should be on at all times. The child may perceive the phone as difficult to use, and practicing the use of the device takes a lot of time.

Though children's use of the mobile phone contains variation, most of the interviewees had adopted a pragmatic attitude to the device. According to Sherry Turkle, "children often use the phrase 'sort of alive' to describe the computer's nature." Even though children nowadays view computers as "just machines," they "continue to attribute psychological properties to them ... that were previously reserved for people." Turkle went on to state that electronic products are "significant actors in provoking a new discourse about aliveness" (Turkle, 1998, pp. 319–320).

Discourse about aliveness and personifying the phone is visible particularly among teenagers, and it is connected to the mobile phone's more significant position in their life. Life may be divided into life before and after acquiring a mobile phone. As an 18-year-old boy put it, "When you've used Nokia mobiles all your mobile life, you're used to their user interface."

8. "TWEENIES" IN MOBILE FEVER

Children start to become interested in mobile communication between the ages of 10 and 12 when their friendship circle expands and hobbies and other activities start to take place outside the home more. Children in this significant phase between childhood and teenage have been called "tweenies." They show signs of mobile fever, but they are also starting to get interested in online chat, music, popular culture, and fashion. Tweenies' mobile culture has its own cultural characteristics, such as teasing by sending empty text messages and developing new variations on "bomb call" games.[1] Although tweenies' use of the mobile phone is increasing, it still remains below that of actual teenagers.

According to City of Helsinki Urban Facts, more girls than boys in this age group have mobile phones. One possible reason is that girls develop more rapidly and are generally more sociable and talkative than boys. Another reason is that mobiles are bought for girls for security reasons more often than for boys.

9. DISCUSSION

So far, research on technology and the information society has concentrated largely on the macro level of the phenomenon: Issues connected to globalization and the economy have formed the core of the research. Empirical everyday information on what the life of children and teenagers is like among the constantly advancing technological developments of the information society remains relatively scarce. This is why knowledge of the micro level from the point of view of young people is increasingly important.

ACKNOWLEDGMENTS

We thank Hanna Liikala for translating the chapter and Leopoldina Fortunati and James Katz for their time and effort in reviewing it.

---

[1] A "bomb call" is made by letting the phone ring for a short time so the recipient does not have time to answer. Because the number of the caller is stored in the phone's memory, the identity of the caller is revealed to the recipient. The number of calls made is shown on the screen of the mobile.

REFERENCES AND BIBLIOGRAPHY

Brosnan, M. J., & Davidson, M. J. (1994). Computerphobia—Is it a particularly female phenomenon? *The Psychologist, 2*(7), 73–78.

Chiaro, M., & Fortunati, L. (1998, August). *New technologies and user skills: A preliminary analysis.* Paper presented at the 14th World Congress of Sociology, Montreal.

Fortunati, L. (1995). *Gli italiani al telefono* [The Italians on the telephone]. Milan: Franco Angeli.

Håpes, T., & Rasmussen, B. (2000a). New technology increasing old inequality? In E. Balka & R. Smith (Eds.), *Women, work and computerization. Charting a course to the future* (pp. 233–240). Boston: Kluwer Academic.

Håpes, T., & Rasmussen, B. (2000b). Young girls on the Internet. In E. Balka & R. Smith (Eds.), *Women, work and computerization. Charting a course to the future* (pp. 241–248). Boston: Kluwer Academic.

Hurlock, E. B. (1978). *Child development* (6th ed.). New York: McGraw-Hill.

Morley, D. (1986). *Family television: Cultural power and domestic leisure.* London: Comedia.

Nurmela, J., Heinonen, R., Ollila, P., & Virtanen, V. (2000). *Mobile phones and computers as parts of everyday life in Finland.* Reviews 5/2000, Statistics Finland.

Oksman, V. (1999). "Että ei niinku tykkää ollenkaan tietokoneista … on vähän niinku outsider": Tyttöjen tulkintoja tietotekniikasta ["When you don't like computers at all … It kind of makes you an outsider": Girls' interpretations of ICTs]. In P. Eriksson & M. Vehviläinen (Eds.), *Tietoyhteiskunta seisakkeella: Teknologia, strategiat ja paikalliset tulkinnat* [Standpoints of information society. Technology, strategies, and local interpretations] (pp. 58–71). Jyväskylä, Finland: SoPhi.

Piaget, J. (1977). *The essential Piaget.* H. E. Gruber & J. J. Vonéche, Eds. New York: Basic Books.

Silverstone, R., & Hirsch, E. (1999). *Consuming technologies: Media and information in domestic spaces.* New York: Routledge.

Statistics Finland. (2001). *Finland in figures.* Retrieved October 2, 2001, from http://tilastokeskus.fi/tk/tp/tasku/taskue_vaesto.html

Turkle, S. (1998). Cyborg babies and cy-dough-plasm. Ideas about self and life in the culture of simulation. In R. Davis-Floyd & J. Dumit (Eds.), *Cyborg babies: From techno-sex to techno-tots* (pp. 67–89). New York: Routledge.

Wood, S. B. (1976). *Children and communication: Verbal and non-verbal language development.* Englewood Cliffs, NJ: Prentice-Hall.

CHAPTER TWELVE

Women's Identities and Everyday Technologies

Raimonda Riccini

The body electric and electrified of Walt Whitman, evoked as an introductory icon by Tomás Maldonado in chapter 2 of this volume, has its female counterpart in the *Eve Future* of Philippe-Auguste-Mathias Villiers de l'Isle Adam (1838–1889), first published in 1886. Whitman's body unequivocally refers to humanity on its adventure in the company of things technical. In the story by the French author, on the other hand, in the literary crevices of an enthusiasm for technology and its creator, we find a disturbing metaphor for the feminine condition.

Let us follow the plot line in the words of the poet Stéphane Mallarmé: "A young lord suffered from the fact that his lover had an imperfection, a certain vulgarity unnoticed by the world but perceived by him ... Edison replaces her with a particularly ingenious robot of his own making" (Villiers de l'Isle-Adam, 1886/1966, p. xiv). Thus, there is a woman, whose body expresses the grandeur of peerless beauty, but whose intelligence fails to satisfy the man who is help-lessly in love with her. There is a man, desperate about the "lack of accord between the physique and the spirit" (Villiers de l'Isle-Adam, 1886/1966, p. 41) of the woman he loves. There is the hero of the tale, the wizard of Menlo Park, Thomas Alva Edison, "who had imprisoned the echo," who had captured energy and channeled it, in luminous rivulets, toward the places where men dwell. Finally, there is technology, which infuses the body of the gorgeous "andreid"[1] with the splendor of intelligence.

Thus a technical feat makes it possible to overcome the limitations—terrifying ones indeed—of the real woman, creating not only physical features but also a personal identity (see the many pages of detailed description of the technology utilized for the construction of the body of Hadaly, in particular Villiers de l'Isle-Adam, 1886/1966, pp. 155–200).

---

[1]In contrast to the usual term *android*, this term represents the "feminized" version of the word.

1. EVERYDAY LIFE AND IDENTITY BASED ON TECHNIQUE

In my contribution I would like to draw attention to the body as the protagonist of everyday life, attempting to identify, at least in broad terms, the relationship between the technology present in the home and the identity of women. I am well aware of the difficulties involved in analyzing this subject, but I believe it is one of the crucial questions for the theme addressed here by scholars of different origins, cultures, and disciplines, especially with the advent of technologies of automation and network telecommunications in the home and their concrete manifestation as domotic habitation models.

From the vast literature that now exists on the intelligent home, my contribution selects a relatively unexplored niche: the changes in the techniques of the female body (Mauss, 1950), as they have been consolidated over the course of a century of mechanization of housework (Giedion, 1967), in the face of the perspectives of the computerized management of work in the home and from outside it. Thus I focus on the body of women as the protagonists—at times in spite of themselves—of everyday technologies.

These are two poles worth defining precisely from the outset. Everyday technologies are seen as the grouping of practices and techniques that form the basis for domestic material culture, with particular reference to the use of instruments, equipment, machines, and systems of objects present in the home (an initial outline of this notion can be found in Esposito, Maldonado, & Riccini, 1981). The female body and identity, on the other hand, are interpreted here in a limited sense, to the extent—exclusively—that they are connected to the segment of everyday life in which women perform the work of a homemaker, within the modern family.

We are well aware of the fact that the theme of identity is closely intertwined with that of the body, with the body *that we are and that we become.* As has often been pointed out, the body is one of the foundations on which identity is built, especially in complex societies. From earliest childhood, the development of bodily identity represents the plot upon which to construct the self as persona, with the multiple identities and roles we then assume over time. But how everyday technologies can act, or interact, with respect to this identity is less clear, especially when real or presumed radical changes are foreseen in domestic organization as the result of the capillary spread of computer and telecommunications technologies inside residences.

Moreover, the practices of everyday life, and especially those that take place inside the home, are located in an opaque zone of research. Domestic activities are a highly decentralized social phenomenon, and therefore represent a field that is difficult to comprehend from many points of view, including those of economics (see Bose, Boreano, & Malloy, 1984; Fortunati, 1981) or historical analysis. (I might also add that "the domestic environment is characterized by a specific static quality with respect to technological innovations; it does not progress rapidly, but remains attached to a series of historical factors, traditions, habits and social norms that make it particularly static and resistant to evolutions"—Diodato, 1996, p. 34). For the most part, household activities take place, to use an evocative image of Mario Praz (1964), in an "intramural landscape," a space that is often intangible, invisible, or not subject to the measurements of the scholar. Access to this space is denied or permitted on a case-by-case basis, and is always regulated by the inhabitant. It is a technical space, it is true, but it is also "emotionally" charged, with personal nooks with which even the homemaker herself may have difficulty coming to terms.

Any outline of the current relationship between technologies and the home, everyday life, and identity encounters an additional difficulty. In contrast to other recent technologies such as the mobile telephone, which have penetrated very rapidly and are already the subject of in-depth sociological analysis (as can be seen in other contributions to this volume), domestic technologies connected to the revolution in telecommunications and information are still in the phase scholars call "negotiation." Bargaining is taking place between the potential of the technologies and the social system, between the supply of innovation and the readiness of the social body to absorb it, between an already defined "function framework" and a "usage framework" that is still problematic and unfocused (on the notion of the frameworks of function and usage, see Flichy, 1995).

Such negotiation takes place whenever a new technology with highly innovative characteristics appears on the social scene. A dialectic is triggered between the social actors and the technology, in a situation of acceptance or rejection, modification and integration. (One of the extraordinarily interesting things about this moment in history is that we, as individuals, are playing an active part in this negotiation, to the extent that we discuss, analyze, and participate as users and consumers, and—at least some of us—as the people who actually design the parts and components involved.) In spite of the limitations indicated by a number of writers (Maldonado, 1998), the constructivist interpretation of innovation makes it possible to avoid the treacherous shoals of technological determinism (Williams, 1998) and even to chip away at the overly rigid positions of the Foucault school regarding domestic technology as a device of social control (Isaac, Fritsch, & Battegay, 1977).

In order to untangle this intricate theme, perhaps it is worth rapidly summing up the principal phases of construction of the relationship between female body/identity and everyday technology, limiting the overview for reasons of space to the three major historical and conceptual passages that have marked its evolution: the *civilization*, the *rationalization*, and the *individualization* of the body. The first primarily involves the processes of socialization; the second involves processes of organization, mechanization, and standardization; the third involves processes of identity. Each of these phases is closely intertwined with the technical and the technological. Each even has a more or less well encoded disciplinary and normative corpus, with manuals of etiquette and manners in the first case, ergonomics or home economics for the second, aesthetic models and fashion canons for the third.

2. SCIENTIFIC MANAGEMENT AS A WAY OF LIFE

I would like to focus, in particular, on the processes of rationalization of the body and the accompanying processes of control. The reasons for this are very simple: The rationalization of the body is "a deeply gendered phenomenon" (Shilling, 1993, p. 38), which lends itself to interpretation from a feminine viewpoint. If dress is one of the primary instruments of the structuring of the female body, domestic technologies and techniques are perhaps of equal or even greater importance. Naturally when we speak of processes of rationalization we are immediately reminded of large-scale production and organizational systems. The application of interpretation models from economic theory, with their focus on the importance of the human–machine relationship for productive efficiency, has accompanied the history of the domestic technological revolution since the late 1800s.

Among the first studies of the impact of technologies on life and work in the home we find those conducted beginning in the mid-1970s by American scholars in the field of the history of technology. The main contributions in this direction remain those of Ruth Schwartz Cowan (1983). In her analysis of the industrial revolution in the home, technology is seen as the element that permits us to understand the *process of housework*, or more precisely the *process of industrialization* of housework. This approach has certainly shed light on many issues, but it has left other equally important questions in the shadows. The analogy between housework and factory work has proven to be quite appropriate, in many ways, but it also brings with it certain preconceptions; the questions raised concerning housework too closely reflect those previously applied to systems of factory work.

In spite of fluctuating and even profoundly ambiguous positions with respect to the technologies of the home, women's thinking has contributed significantly to reinforce and sustain the productivist analogy. This may explain, at least in part, the growing attention women have paid to "modern" systems and devices for correct management of the home, and the importance attributed to mechanization of operative processes and the role of appliances in easing the labors of the homemaker. At first the identity image was based on a model we could define as feminist–emancipationist, aimed at reducing the labor of women in daily practical matters. Soon, however, this image was replaced by one closer to the Taylorist focus on enhanced efficiency. In the face of growing technical complexity inside the home, the study of single operative processes appeared to be utterly insufficient. A need was felt to move on to the analysis of the entire process, including different operative procedures that were increasingly interdependent. The analogy of the model of the factory—a Fordist factory, at this point—permitted analysis of the organization of housework mechanized as a system regulated by laws and measurable in its space–time components, as theorized by Taylor. As Ellen Richards put it, "The work of homemaking in this scientific age must be worked out on engineering principles and with the cooperation of trained men and trained women. Tomorrow, if not today, the woman who is to be really mistress of her house must be an engineer, so far as to be able to understand the use of machines" (quoted in Cowan, 1990, p. 149). As we can see, the theorizing hypothesized the "scientific performance of housework" as a veritable ideology. This was a real movement for the science of domesticity, with the initial participation, above all, of Americans such as Charlotte Perkins Gilman, Lilian Moller Gilbreth, Christine Frederick, Mary Pattison, and Ellen Richards, but then also spreading rapidly in Europe, with figures such as Paulette Bernège. This ideology was spread by instruction on the new technologies filtered by what Alain Corbin called the "somatic disciplines" (Corbin, 1994, p. 483) home economics schools, specialized magazines, women's periodicals, initiatives by agencies and institutions, commercial advertising, and, finally, architectural manuals. Similarly to what happened for the nonspecialized worker in the Fordist factory, all this was oriented toward simplifying and rationalizing tools and procedures, to make sure that the work could initially be performed by a "relatively unskilled worker" (Banta, 1993, p. 241). But there was more. Scientific management contributed in the formation of a detailed system of values and preferences, recommendations and abilities, through the popularization of a series of norms of behavior, gesture, and even ritual, all based on efficiency, control, and uniformity.

From our point of view it is important to emphasize the fact that the main connotation of this propaganda was an attempt to relate (or perhaps to subject) the *techniques of the body* (the notion introduced by Mauss, 1950) related to domestic activities to the appearance in the home of systems, utensils, tools, machines, furnishings, and foods—all new and in constant growth. The techniques of the body, which according to Marcel Mauss are not necessarily tied to instruments but instead based on the body itself, become increasingly dependent on the artificial equipment of the micro-environment. They are forced to come to terms with the growing presence of artifacts and activities connected to artifacts that are gradually more and more specialized and differentiated.

But, again according to Mauss, technique is an effective act only when it is a traditional act, because it is based on transmission. And no technique or transmission can exist without *tradition*. In the preindustrial world, tradition was guaranteed by the learning of techniques in the family: Consider, in particular, the care of children, culinary practices, the washing of clothing. As this system of transmission gradually broke down, especially as a result of the fragmentation of the extended family, other forms of acculturation came into play. They included home economics manuals, which became the basis of a new tradition and the vehicle for its transmission, precisely because they provided the codes for the new techniques of the body, linked to new instruments—effective ones, but no longer traditional. This form of instruction had a function of "substitution," so to speak.

Moreover, home economics and the other somatic disciplines favored the passage from the oral tradition of domestic know-how—transmitted in a continuous hand-to-hand relationship with other women—to a written, iconic culture. This process has been radicalized, shattering into forms of learning that now happen in a hand-to-hand relationship with technologies, artifacts, and media. Just consider the educational and informational function of television, or the many forms of transmission of domestic knowledge that take place through the purchasing, consumption, and use of products. These forms can be directly expressed through products (e.g., in recommendations for use, modes and schedules of conservation, or the functional and procedural indications provided on the control panel displays of home appliances), but also through the technologies themselves (washing machines that automatically analyze the type of fabric and the level of soiling, electronic ovens that propose dietetic recipes, food selection performed by smart refrigerators).

In conclusion, domestic education has played a decisive role in the processes of assimilation of technology in the more general context of the development of cultural and material models of living. Roughly speaking there has been a passage from a first phase in which the main task of this type of material was to give an identity to the figure of the housewife—in its practical aspects, but also and perhaps principally in moral terms—to a second phase in which women are made familiar with the great technological innovations, especially those linked to the use of gas and electricity. Finally, the material promoted all the devices designed to facilitate housework, with emphasis on the role of electric home appliances.

It would be restrictive to limit the function of this literature only to the legitimization of the new technologies. In effect, it played a larger role by defining, through the figure of the housewife, a sort of new identity statute. This contributed to form a typology of the rational, competent, modern woman—the same woman who, having learned to make the best use of technology, would thus be ready to become a *consumer*.

In this sense, Taylorization is completely consistent with the ideology of the society of consumption. As we all know, by freeing up time for oneself through the use of labor-saving devices, women are enabled (or should be enabled) to devote more time to other activities, including work outside the home. This was the beginning of the vicious circle in which the freeing of time in itself becomes the availability of time to produce the buying power required to purchase products for the freeing of more time.

Today we are aware of the fact that this prospect has been transformed into a social paradox.

3. DOMOTICS AND AUTOMATED LIFESTYLE
(OR THE BODY SYSTEM AS A PART OF LIFESTYLE)

To continue—with all due caution—with the productivist analogy, we can say that today we are faced with a situation in which housework can be seen in a "post-Fordist" perspective: "The obsession with standardization that had functioned as a catalyst for energy is replaced today by the obsessive race to de-standardize everything: tasks and products.— The specialization of the Fordist world is replaced by the 'multivalence' of our modern times" (Cohen, 2001, pp. 25–26).

In place of the concept of order, organization, and saving of time and effort, contemporary technologies tend to propose other values and other identity profiles. The dwelling-machine, with more or less rigidly defined spatial and functional parameters, is replaced by a "fluid" home, interwoven with a "nervous system" of technical systems and connections to bring functions to the user at will, accentuating the communicative aspects. The model of this type of dwelling is the so-called "smart house."

Though it is a familiar term now, even included in commercial slogans, it is still worth reviewing the concept of domotics along essential lines. Domotics is a programmable centralized system that regulates all the technological functions of the home. "The entire home is seen as an organic complex of structures and services, integrated in a rational, efficient manner to respond to specific usage requirements" (Enea, 2001).

Today the market offers a wide range of domestic devices for control, monitoring, security, and telecommunications: microcontrollers (control devices for electronic machines), technologies connected with robotics (distance sensors such as infrared rays or ultrasound, or position sensors, for the activation and management of home appliances), measurement sensors (e.g., for monitoring air quality), high-efficiency activators and electronic motors (Diodato, 1996, p. 38). In practice, although these electronic devices are available, and at a feasible cost,[2] we are still not capable of applying them in a simple way in functionally integrated systems.

Nevertheless the system of integrated control of all the technological equipment in the home is the precise basis for all the experimentation conducted in this field around the world, starting in the 1980s: management and control functions (energy: heating and lighting; security: burglar and fire alarms; communication: inside the home and between the home and the outside world; entertainment: television, CD, video, computer games; Internet; environment: temperature, indoor pollution) concentrated in centralized control panels. In the United States, the first long-term global project involving large companies in the fields of electronics and telecommunications, in collaboration with associations of contractors and real estate firms as well as public and private research institutes, was that of the "Smart House" in 1984. In Europe, the first conference on the intelligent habitat, organized by the Association for the Home of the Future and the sector associations of the French construction, electronics, and home appliances industries, in collaboration with the European Community, was held in Paris in 1988 (Guidi, 1988). This was the origin of the Esprit research project.

Although applications results on a wide scale are still not available, it is already possible to identify some relevant questions regarding this theme. The first and most evident effect of the integration of domestic technology systems is the great

---

[2]"The cost of an intelligent system for the average home, including design, installation, programming and testing, ranges from 10 to 15 million lire [about U.S. $5,000–7,500], depending upon the number of connected devices and the type of applications involved (burglar alarms, environmental safety, heating and air conditioning control, control of lighting and home appliances, blinds, etc.)" (Mongiovì, 2001).

emphasis placed on the interactive and communicative aspects of the objects and systems involved. In this perspective the direct relationship with single pieces of equipment vanishes. Mechanical ergonomics is replaced, so to speak, by an ergonomics of communication, now the focus of the most advanced research in the field of cognitive ergonomics. Rather than a relationship between the user and individual technologies, we can observe a relationship with the entire environment created by information and communication technologies. These permit people to operate and communicate with machines, and to communicate with each other through machines (Mantovani, 1995). This change should also enable greater flexibility in working life as a whole: Segments of nondomotic work can be performed in the home (telework), as can segments of house-related work usually performed outside the home (shopping, banking, even entertainment functions).

This brief description suffices to indicate how domotic technologies can give form not so much (or not only) to specific activities, as to the definition of a new lifestyle, in which terms such as comfort, free time, relaxation, well-being, and security predominate, and in which housework would be dematerialized, impalpable. Once again, as in the 1800s, the function of this dwelling would be simply "regenerative" (Esposito et al., 1981).

Moreover, what seems to be on the horizon is a drastic revision of the role of the human body, especially the female body and its techniques, in the activities of work and of housework in particular. The regimen of the body—apparently liberated from the burdens of physical work—becomes part of a new lifestyle in which great emphasis is placed on the growth of knowledge (Shilling, 1993). Extending the analogy with the productive world, "one of the fundamental traits of post-Fordist working processes, or the breakdown of the rigidly end-oriented connotations of work, is the spread of 'feminine' models of production, in which the immaterial nature of the result is predominant, as therefore are the elements connected with knowledge, language, assistance, care" (Zanini & Fadini, 2001, p. 61).

Without wishing to underestimate the potentially positive aspects of this scenario, the suspicion—or something more than just a suspicion—arises that once again, disguised behind utopian prospects, we are being persuaded to accept new forms of subservience to housework. What social design is concealed behind this image? What is the inherent project in domotic technology? What is the role of women in this project?

At present the answers to these questions are very problematic. As Anne-Jorunn Berg has demonstrated, the central importance of housework in this new dwelling, the work effected inside the "new domestic landscape," is in practice overlooked. If we analyze the projects developed in different parts of the world, we cannot help but notice that the smart home is, once again, a masculine technical construction, in which the functions of the technology are determined by the designers and the producers. "The smart house is a typical case of technology push as opposed to consumer pull. Its conception is linked not to the practices of everyday life, but to a fascination with what is technically possible" (Berg, 1995, p. 85). It also reflects a male idea of the home and responds for the most part to the activities men perform there; it contains no significant elements to replace or reduce housework. At best, the work can be shifted to different moments, permitting optimization of productive time both outside and inside the home. The cyborg-homemaker is invited to partake of this "technological banquet" (Haraway, 1995), but she will still have to clear the table and straighten up the house.

REFERENCES AND BIBLIOGRAPHY

Allen, B., Berlo, A., & Ekberg, J. (2001). Design guidelines on smart home. Retrieved May 14, 2001, from http://www.smart-homes.nl

Attfield, J., & Kirkham, P. (Eds.). (1989). A view from the interior. Feminism, women and design. London: The Women's Press.

Banta, M. (1993). Taylored lives. Narrative productions in the age of Taylor, Veblen and Ford. Chicago: University of Chicago Press.

Berg, A.-J. (1995). A gendered socio-technical construction: The smart house. In N. Heap, R. Thomas, G. Einom, R. Mason, & H. MacKay (Eds.), Information technology and society (pp. 74–89). London: Sage.

Berg, A.-J., & Lie, M. (1995). Feminism and constructivism: Do artifacts have gender? Science, Technology, & Human Value, 20(3), 332–351.

Bernège, P. (1933). Quand une femme construit sa cuisine [When a woman constructs the cuisine]. Reprinted in Culture Technique, no. 3, September 1980.

Bettinelli, E. (1991). La prossima casa. Progetto e tecnologia della automazione domestica [The next house. Project and technology of domestic automation]. Milan: Idea Books.

Bose, C. E., Boreano, P. L., & Malloy, M. (1984). Household technology and social construction of housework. Technology and Culture, 47(4), 725–748.

Caronia, A. (1981). Abitanti dell'utopia telematica [Dwellers of telematic utopia]. Alfabeta, 28, 10.

Cohen, D. (2001). I nostri tempi moderni. Dal capitale finanziario al capitale umano [Modern times. From financial to human capital]. Turin, Italy: Einaudi.

Corbin, A. (1994). Dientro le quinte [In the wings]. In P. Ariès & G. Duby (Eds.), La vita privata. L'Ottocento [Private life. The 19th century] (pp. 332–486). Milan: Mondadori.

Cowan, R. S. (1983). More work for mother. The ironies of household technology from the open hearth to the microwave. New York: Basic Books.

Cowan, R. S. (1990). Ellen Swallow Richards: Technology and women. In C. W. Pursell, Jr. (Ed.), Technology in America. A history of individuals and ideas (pp. 142–150). Cambridge, MA: MIT Press.

Diodato, C. (1996). La casa del futuro [The house of the future]. In G. Triani (Ed.), Casa e supermercato. Luoghi e comportamenti del consumo [House and supermarket. Consumer places and behavior]. Milan: Elèuthera.

Elias, N. (1982). Il processo di civilizzazione [The process of civilization]. Bologna: Il Mulino.

Enea (Ente per le Nuove Tecnologie, l'Energia e l'Ambiente). (2001). Introduzione al concetto di "smart house" [Introduction to the concept of the "smart house"]. Retrieved March 14, 2001, from http://andi.casaccia.enea.it/andi/tecno/smart/smh1co.htm

Esposito, S., Maldonado, T., & Riccini, R. (1981). Condizione femminile e ideologia del comfort [The female condition and the ideology of comfort]. Casabella, 45(467), 27–33.

Flichy, P. (1995). L'innovazione tecnologica [Technological innovation]. Milan: Feltrinelli.

Forester, T. (1991). The myth of the electronic cottage. In T. Forester (Ed.), Computers in the human context. Information technology, productivity, and people (2nd ed., pp. 213–227). Cambridge, MA: MIT Press.

Fortunati, L. (1981). L'arcano della riproduzione. Casalinghe, prostitute, operai e capitale [The arcana of reproduction. Housewives, prostitutes, female workers and capital]. Venice: Marsilio.

Fortunati, L. (Ed.). (1998). *Telecomunicando in Europa* [Telecommunicating in Europe]. Milan: Franco Angeli.

Giedion, S. (1967). *L'era della meccanizzazione* [The age of mechanization]. Milan: Feltrinelli.

Gras, A., & Moricot, C. (Eds.). (1992). *Technologies du quotidien. La complainte du progrès* [Sociology of everyday technology]. Paris: Editions Autrement.

Gras, A., Joerges, B., & Scardigli, V. (Eds.). (1992). *Sociologie des techniques de la vie quotidienne* [Daily technologies. The complaint of progress]. Paris: L'Harmattan.

Guidi, P. (1988). La casa elettronica: un futuro comfort sotto controllo [The electronic house: Future comfort under control]. *Apparecchi elettrodomestici nella casa moderna* [Household appliances in the modern house], 36(1), 78.

Haraway, D. J. (1995). *Manifesto Cyborg. Donne, tecnologie e biopolitiche del corpo* [Cyborg manifesto. Women, technologies and bio-politics of the body]. Milan: Interzone Feltrinelli.

Isaac, J., Fritsch, P., & Battegay, A. (1977). Discipline à domicile. L'èdification de la famille [Home discipline. Family education]. *Recherches*, no. 28, November.

Maldonado, T. (1997). *Critica della ragione informatica* [Critique of IT reason]. Milan: Feltrinelli.

Maldonado, T. (1998). Ancora la tecnica. Un "tour d'horizon" [Technology again. A survey]. In M. Nacci (Ed.), *Oggetti d'uso quotidiano. Rivoluzioni tecnologiche nella vita d'oggi* [Objects of daily use. Technological revolutions in everyday life today] (pp. 197–227). Venice: Marsilio.

Mantovani, G. (1995). *Comunicazione e identità. Dalle situazioni quotidiane agli ambienti virtuali* [Communication and identity. From everyday situations to virtual environments]. Bologna: Il Mulino.

Mauss, M. (1950). *Sociologie et anthropologie* [Sociology and anthropology]. Paris: Presses Universitaires de France.

Mongiovì, P. (2001). *La casa intelligente* [The smart home]. Retrieved March 14, 2001, from http://www.initalia.it/sistemacasa/casintelligente.htm.

Moses, D. (1997). *Automated lifestyle.* Retrieved March 14, 2001, from http://www.hometoys.com/htinews/dec97/articles/moses/moses.htm

Paoni, S. (1997). Smart House ovvero ... casa intelligente [Smart house or casa intelligente]. *B.I.T. on line*, 2(2). http://www.ptu.sitech.it/bit/1997

Popitz, H. (1996). *Verso una società artificiale* [Towards an artificial society]. Rome: Editori Riuniti.

Praz, M. (1964). *Filosofia dell'arredamento* [The philosophy of furniture]. Milan: Longanesi.

Rendell, J., Penner, B., & Borden, L. (Eds.). (2000). *Gender, space, architecture. An interdisciplinary introduction.* London: Routledge.

Robinson, K., & Webster, F. (1988). Cybernetic capitalism: Information, technology, everyday life. In V. Mosco & J. Wasko (Eds.), *The political economy of information* (pp. 44–75). Madison: University of Wisconsin Press.

Shilling, C. (1993). *The body and social theory.* London: Sage.

Stone, D. (1997). Work and the moral woman. *The American Prospect*, November–December, 78–86. http://epn.org/prospect/35/35stonfs.html

Villiers de l'Isle-Adam, P.-A.-M. (1966). *Eva Futura* [The eve of the future]. Milan: Bompiani. (Original work published in French as *L'Eve future*, 1886)

Weisman, L. K. (1994). *Discrimination by design. A feminist critique of the man-made environment.* Urbana: University of Illinois Press.

Williams, R. (1998). The political and feminist dimensions of technological determinism. In M. Roe Smith & L. Marx (Eds.), *Does technology drive history? The dilemma of technological determinism* (pp. 217–236). Cambridge, MA: MIT Press.

Zanini, A., & Fadini, U. (Eds.). (2001). *Lessico postfordista. Dizionario delle idee di mutazione* [Post-fordist lexicon. Dictionary of ideas of change]. Milan: Interzone Feltrinelli.

Zuboff, S. (1984). *In the age of the smart machine. The future of work and power.* New York: Basic Books.

CHAPTER THIRTEEN

Body to Body:
Copresence in Communication

Alberta Contarello

People who talk may be communicative and construct together possible worlds oriented toward promoting reciprocal abilities. Or they may be confrontational—interfering, inventing sanctions, exchanging pain and suffering. These "realities" are central topics in the social sciences—particularly in social psychology, communication research, and studies on personal relations, whose aim is to find out to what extent the production or modification of the social psychological world is due to individuals, and to what extent it is due to their relationships.

Several studies have stressed the importance of relationships not only in an individual's formative phases but across the whole lifespan. And if relationships play a foreground role, how can we distinguish between direct and mediate relationships? What effects can be foreseen for the ever-growing communicative potential that new technologies offer to individuals nowadays? Different discourses intertwine with the aim of answering such questions. Some underline the "naturalness" and thus the strength of relations free from any sort of artifice. Others stress the symbolic valence and thus the power of relations that, with the help of prosthetic tools of various kinds, would enable us to transcend the space and time dimensions. Others note serious shortcomings in both of these perspectives, but fear a worsening of the relational world as a result of the development of new techniques and technologies that make face-to-face relations less and less necessary.

The aim of this chapter is to reflect on some theoretical and empirical approaches that, though enhancing the role of social relations in the construction and communication of meaning, focus on the role of the body and of copresence in interactions and can thus contribute to understanding recent changes brought about by the introduction of new technologies. Three main sources are considered: the historically important thinking of G. H. Mead; empirical evidence from the "nonverbal" literature of the second half of the 20th century; and recent research, from different disciplinary perspectives, that focuses on the role of the body and of its presence—real and symbolic—in social and psychological processes.

1. GESTURE

George Herbert Mead maintained that "mind arises through communication by a conversation of gestures in a social process or context of experience—not communication through minds" (Mead, 1934, p. 50). Key concepts in Mead's theorizing are the construction of meaning and of self from gestures, the role of the hands (mainly by virtue of the opposing thumb) in enabling humans to develop and use tools, and the necessity of taking one another's perspective in order to convey meaning and reach mutual understanding. These give the human body's physiology and its expression a central role.

Mead's thinking is known in the social sciences mainly for its part in giving birth to symbolic interactionism and, more recently, to social constructionist models, which have made an important contribution to understanding changes in our cultures concerning the massive intervention of new technologies (Gergen, 1991). It is the subject of renewed interest (Farr, 1996) mainly because of the potential knowledge and interpretation it offers to contemporary scholars facing sophisticated tools of high symbolic value. It is rooted, in fact, in a "conceptual tripod" that draws on Darwin's thinking, on a social behaviorism that differs from Watson's in putting act and social act at the core of theoretical thinking, and on a deep focus on the individual's viewpoint, which recalls Einstein's theory of relativity.

A theory of its own time, some might say. To what extent and to what degree can it help us to understand new phenomena, such as the recent opportunities to communicate in different times and places, that were virtually impossible to foresee in the past century and millennium?

The basic idea is that the processes underlying the construction of meaning, as well as of self and identity, are social processes; they are, by definition, not "natural" but relational ones. The importance of gestures and of play—mainly role playing—emphasizes the need for "social rehearsals," for coordinating practices that contribute to the construction of individuals. Such exercises are fundamental, both in the developmental phases of the individual and throughout the lifespan. The capacities of symbolic mediation increase with the growth of the individual and the group, but the strength of gesture coordination and the interacting dance that involves bodies as well as minds extend over time. Many decades later, in line with Mead's observations and in parallel with Vygotskij's, Bruner (1975) showed how turn taking between the child and his or her caregiver is a prerequisite to language through the use of formats or routines. The latter underlie the role of social coordination in the acquisition of sensorimotor schemas as well as of language, and most probably derive from the dance between action and rest that binds mother and child in the feeding process. Studies by Condon and Sander (1974), moreover, showed how newborns only a few days old synchronize their movements with the sound of a human voice (even if the language is different from that of the babies' parents) but not with sounds unconnected to language. The importance of such coordination is not limited to initiating the most distinctive of human potentials, but continues throughout the lifespan (see Kendon, 1970).

2. TURNS

After a long period of silence, attention was again focused on the body with the nonverbal "turn," which was consolidated mainly between the 1960s and 1980s. From a different perspective, it underlined the importance of interpersonal coordination, focusing attention on the body and its communicative potency and peculiarities. A corpus of research developed, aimed at furthering our understanding of nonverbal aspects of communication. Eye contact, facial expression, head movement, touch, gestures, postures—all were submitted to painstaking analysis in order to discover their functioning and role within social exchanges (see Argyle, 1975 / 1992).

It was seen that verbal and nonverbal cues combine harmoniously to achieve and maintain a balance between opposing forces as regards the optimal interpersonal distance. In this vein, Argyle and Dean (1965) proposed their equilibrium theory: eye contact, distance, and the intimacy of a conversation act in concert to achieve the desired level of psychological closeness, which is often negotiated dynamically by the interactants.

It was also seen that the use of personal space assumes intense psychological meaning and is modulated according to the participants' culture. Hall (1966) distinguished various levels of distance between people—close, personal, social, public—and suggested an important role for culture in the quantification of such distances. Systematic observations, aimed at supporting such ideas with empirical evidence, have, however, stressed the role of the body and the scope of the five senses in the definition of "presence" and "co-presence" (Ciolek, 1982). The way in which people become aware of the presence of another person is affected by ecological and cultural factors that define different kinds of social situations and spatial arrangements and do not correspond exactly to the physical distance between people. Such a perception, however, appears to be strongly guided by sensory data: Sight, hearing, smell, and touch intervene, in different ways and to different extents, in the identification of fields of copresence (Ciolek, 1980). One might conclude that culture and shared norms contribute to define and identify psychological experiences and "right" ways of using interpersonal space, but that the degree of freedom of social conventions is linked to our biological features as social animals.

Of special interest in this context are studies on interpersonal coordination, which suggest that mental tuning (the congruence of mental states) is reflected in the congruence of physical behavior between interactants. Matching, especially the mirroring of postures or the "postural echo" (Morris, 1977), creates a space for shared viewpoints, contributing to a feeling of "togetherness," harmony, and being "in tune" (Scheflen, 1964). Studies in interpersonal coordination, behavior matching, and interactional synchrony show that these phenomena, although quite complex and not completely understood, are very important for effective interpersonal relations and rapport (Bernieri & Rosenthal, 1991).

In a research program to measure and understand such phenomena, Rosenthal and others distinguished between matching—the postural congruence of people, as both imitation and mirroring—and interactive synchrony, which expresses itself through the rhythm of interactions, simultaneity of movements, and meshing of behavior. The process of synchronization is the one that is most closely related to the quality of the rapport between the persons involved. The authors' studies are based on a research paradigm that uses observers' judgments on the four aspects of interaction—posture matching and the three forms of interactive synchrony—in "genuine" and "reconstructed" interactions. In this way, they studied mother–child and teacher–pupil relations, and concluded that, although the reason for interpersonal coordination and the quality of rapport is not yet known, it is clear that "social and emotional relationships are closely linked to a physical aspect of human interaction of which we are usually unaware" (Bernieri & Rosenthal, 1991, p. 429). The study of these relations is worthy of much wider consideration, particularly if we take into account their importance "especially for applied areas of social interaction such as medicine, psychotherapy, education, management, and social relation generally" (Bernieri & Rosenthal, 1991, p. 30).

Clues and suggestions that help us to understand come from different disciplinary perspectives. In a medical psychiatric framework, Fish, Frey, and Hisbrunner (1983), for instance, observed how postural changes by the therapist somehow anticipate the recovery process of severely depressed patients. In a phenomenologically oriented psychoanalytic framework, Resnik (1986) guided us straight to the heart of the question:

> The relation which develops in the field is first of all a situation of co-presence: the body-person of each other influences, affects and conditions the ground of transfer, stimulates the field or, on the contrary, inhibits it, or even, in extreme situations, blocks the Umgang, the ability to exchange, blocks the fluidity of the encounter. (p. 39)

Furthering our knowledge of the role of the body within social relations by drawing on different disciplinary sources also brings new risks and challenges. In the theoretical context to which Resnik referred, for instance, body image and body schema are not somatic but mental phenomena. Nevertheless, this conception, in line with the importance given by Rosenthal et al. to the direct perception of the property of stimuli in the environment ("affordances" in Gibson's terminology; see also note 1 on page 75), might be of help in analyzing the new relationship between the human body and technology.

3. TECHNOLOGY

The rapidly developing new technologies—the Internet, chat lines, video-conferencing, virtual reality—offer powerful tools for the acquisition and dissemination of information and experience. They also open up new and exciting avenues for research. Given the fundamental role of the body in interpersonal exchange and social relations, however, there are reasons to be concerned about the disembodied nature of the communication that the electronic media have created.

It is difficult to document precisely the extent of the phenomenon, which appears to be constantly expanding. In Italy, the telephone, especially the mobile phone, has had remarkable success: ISTAT data for June 2000 show that 48.3 percent of Italian families owned at least one. A similar trend characterizes the Internet and the Web (Fortunati, 1995, 1998). In this case, the data are less reliable: In March 2000 the newspaper *Il Sole 24 Ore* estimated 7–8 million Internet visitors over the previous 3 years; *Panorama-on-line* in November 1999 estimated 5 million (or 9 percent of the population); a Eurisko survey in November 2000 estimated 9.5 million (or nearly 20 percent of the adult population). From many sides, however, we have indications of an exponential growth in the use of old and new technologies that are enabling people to communicate and interact at a distance and that are leading us to redefine some key words in social psychology research. One of these is the concept of "presence."

What, for instance, is the role of the human body in virtual environments? How can we define presence and copresence within them? How can we cope with the absence of sensory information, whose importance in "real" relations, as we have seen, we are far from fully understanding? Both theoretical and empirical research are active in this regard. On the one hand, social constructionism, in its most clear-cut versions, rejects a sharp opposition between "real" and "virtual" or "artificial." Mantovani and Riva (1999), considering that "there are no 'natural', unmediated experiences of presence in an environment," emphasized "the social and cultural dimension of experience" (pp. 543, 545). On the other hand, studies are being devoted to what chiefly characterizes the experience of presence in virtual environments and what role body movements have within them (Slater & Steed, 2000; see also Short, Williams, & Christie, 1976; Steuer, 1995).

Another field of study that new technologies may help to refine is related to "social support." Evidence comes not only from social psychology but also from psychophysiology, neurobiology, and psychoneuroimmunology of the important role played by (positive) body-to-body relations in the promotion and maintenance of both physical and psychological health and well-being.

A review by Cacioppo et al., for instance, demonstrated the relationship between social support and physiological processes and searched for the underlying mechanisms and implications for health. The authors brought together various studies that analyzed the covariation between social support, on the one hand, and aspects of the cardiovascular, endocrine, and immune systems, on the other.

Although the results should be treated with caution on theoretical and methodological grounds, they point to a positive effect on health, mainly in the long run, of positive social relations. The social support offered by such relations is conceptualized and operationalized in various ways—in "structural" terms, through the facilitation of social bonds "in the flesh," and in "functional" terms, through giving emotional and informational support (Uchino, Cacioppo, & Kiecolt-Glaser, 1996). Social support is another concept that might gain clarification from a comparison between the help individuals may offer with their bodily presence and the help they may give on the emotional and/or informational side through telecommunication.

Such clarifications are missing in the contemporary outline of studies, and the challenge of new technologies might offer an extraordinary opportunity to improve our knowledge. The positive effects of (positive) body-to-body interactions are too important to be marginalized with respect to the more unbound forms of interaction (in the fields of friendship, work, and play) promised by the virtual world.

Studies on attachment bonds in children and adults (Bowlby, 1969; Hazan & Shaver, 1994), on emotions (which are closely linked to social contexts and shared norms, but remain "body matters"—Frijda, 1986), on communication between partners and between parents and children (Gottman, 2001; Karney & Bradbury, 1995), on the role of positive social interactions and, especially, of touch,[1] all contribute to form an ongoing research project on the links between "embodied" interpersonal bonds and health (Ryff & Singer, 2000). Further research is thus highly desirable to document more thoroughly such effects and to clarify the similarities and differences between body-to-body communication and communication through technology.

Research on the effects of such new communicative forms is growing, and it has given rise to some lively debates. In particular, attention is devoted to effects on self-perception and interpersonal relationships as well as transpersonal forms of consciousness (Gackenbach, 1998). Also under examination are the effects upon identities (Mantovani, 1995) and personal relationships (Paccagnella, 2000), the effects—facilitating versus inhibiting, short term versus long term—of the electronic media on various forms of interaction, and the social influence exerted by the new media on people and on social contexts (Contarello & Mazzara, 2000).

In a review on the topic, Wallace (1999) took an open position in terms of interpretation. Considering communication via computer in particular—through email, chat lines, and virtual games—Wallace noted that to be able to choose different identities, totally unconnected to one's own individual body, may sometimes take on highly positive features, mainly in cases of communicative or relational impairments. It may also, however, precipitate potentially criti-

---

[1]See studies by Uvmas-Moderg (1998) on the neuro-endocrine correlates of somatosensory stimulation, more specifically on the rise of oxytocin levels linked to such stimulation and related potent antistress effects (see also Field, 1998).

cal relationships into dead ends and exacerbate "natural" aggressive or irresponsible tendencies.

As Wallace noted, the Internet and the Web may offer powerful experimental frameworks for refining theoretical models proposed by social psychological research. The central themes that Wallace reviewed, searching for their peculiarities in the context of telecommunication, are person perception, group dynamics, aggression, prosocial behavior, the development of close relationships, as well as the psychological aspects of pornography on-line and the danger of Internet addiction. Studying the effects of absence—or of limitation of presence—in such contexts might be of great help.

Spencer (2000), for his part, invited us to consider the beneficial effects on health and well-being of interpersonal links on a larger scale. The author is the founder of a nonprofit association to promote live arts experiences, and he maintains that they have a strong positive impact on those who take part in them. Through copresence, features such as intimacy, connection, and the sharing of emotion come to the fore. Furthermore, live arts experiences are unique; the people who take part in them are immersed in uncertainty and unpredictability, while at the same time contributing to and participating in the emotional climate of the event, with a wide range of multisensory information. This, Spencer noted, characterizes and differentiates live art and mediated art and led him to conclude that old and new technology—and particularly the "Internet—is a magnificent tool for acquisition and dissemination of information, but not as a substitute in the arts for live arts experiences" (Spencer, 2000, p. 3).

As we have seen, many questions are being researched, but they all turn on a fundamental issue, which is only apparently new: Are we facing an epochal change that will increasingly replace body-to-body communication with long-distance communication mediated by newer and newer technologies? If this is the case, will we free ourselves of the constraints of the body or will we encounter an impoverishment of the relational strategies and competence that, through time and practice and social rehearsal, face-to-face relations provide?

New technologies are today arriving on a scene that is very rich in symbolic and abstract elaboration and that has taught us to take into account the reciprocal and constructive links between the relational and the cultural context—produced mainly through the media—and characteristics specific to "human nature." But we also know that the body, because of its expressive and communicative strength, can send and receive signals, communications, and clues in a dynamic—and unique—way at different levels and through various channels. We know its power, but only in part its workings. To concede an ever-growing share of our own communicative channels to mediated communication involves serious risks both in terms of psychophysical well-being and as regards forms of social organization that promote and make real the chances of communicative exchange at different levels (see Fortunati, 2001; see also Maldonado, chap. 2, this volume).

Searching for answers to the questions posed by new technologies, the multi-disciplinary research developed in the past few years seems to be able to bring the human body back to the center of attention and to reintroduce onto the agenda questions that have always been part of the story of social psychology. On theoretical and practical grounds, the aim is, on the one hand, to highlight the limits and risks that these tools involve and, on the other, to better understand their potential to bring to the foreground relations that act on multiple registers, through a wider awareness and an enlarging of the range of action of people and their networks of reciprocal bonds.

ACKNOWLEDGMENTS

I thank Peter Collett for his valuable help and stimulating conversations.

REFERENCES AND BIBLIOGRAPHY

Argyle, M. (1992). Bodily communication (2nd ed.). London: Methuen. (Original work published 1975)

Argyle, M., & Dean, J. (1965). Eye-contact, distance and affiliation. Sociometry, 28, 289–304.

Bernieri, F. J., & Rosenthal, R. (1991). Interpersonal coordination: Behavior matching and interactional synchrony. In R. S. Feldman & B. Rimé (Eds.), Fundamentals of nonverbal behavior (pp. 401–432). Cambridge, England: Cambridge University Press.

Bowlby, J. (1969). Attachment and loss. New York: Basic Books.

Bruner, J. (1975). The ontogenesis of speech acts. Journal of Child Language, 2, 1–19.

Ciolek, M. T. (1980). Spatial extent of the field of co-presence: A summary of findings. Man–Environment Systems, 10, 57–62.

Ciolek, M. T. (1982). Zones of co-presence in face-to-face interaction: Some observational data. Man–Environment Systems, 12, 223–242.

Condon, W. S., & Sander, L. W. (1974). Synchrony demonstrated between movements of neonate and adult speech. Child Development, 45, 456–462.

Contarello, A., & Mazzara, B. M. (2000). Le dimensioni sociali dei processi psicologici [The social dimension of psychological processes]. Rome/Bari: Laterza.

Farr, R. (1996). The roots of modern social psychology. Oxford, England: Blackwell.

Field, T. M. (1998). Massage therapy effects. American Psychologist, 53, 1270–1281.

Fish, H. U., Frey, S., & Hisbrunner, H. P. (1983). Analysing nonverbal behavior in depression. Journal of Abnormal Psychology, 92, 307–318.

Fortunati, L. (Ed.). (1995). Gli italiani al telefono [The Italians on the telephone]. Milan: Angeli.

Fortunati, L. (Ed.). (1998). Telecomunicando in Europa [Telecommunicating in Europe]. Milan: Angeli.

Fortunati, L. (2001). La comunicazione tra naturale e artificiale [Communication between the natural and the artificial]. Manuscript submitted for publication.

Frijda, N. (1986). The emotions. Cambridge, England: Cambridge University Press.

Gackenbach, J. (Ed.). (1998). Psychology and the Internet: Intrapersonal, interpersonal and transpersonal implications. New York: Academic Press.

Gergen, K. (1991). The saturated self. New York: Basic Books.

Gottman, J. M. (2001). Meta-emotion, children's emotional intelligence, and buffering children from marital conflict. In C. D. Ryff & B. Singer (Eds.), *Emotion, social relationships and health* (pp. 23–40). Oxford, England: Oxford University Press.

Hall, E. T. (1966). *The hidden dimension.* New York: Doubleday.

Hazan, C., & Shaver, P. R. (1994). Attachment as an organization framework for research in close relationships. *Psychological Inquiry, 5,* 1–22.

Karney, B., & Bradbury, T. N. (1995). The longitudinal course of marital quality and stability: A review of theory, method, and research. *Psychological Bulletin, 118,* 3–34.

Kendon, A. (1970). Movement coordination in social interaction: Some examples described. *Acta Psychologica, 32,* 1–25.

Mantovani, G. (1995). *Comunicazione e identità. Dalle situazioni quotidiane agli ambienti virtuali* [Communication and identity: From the daily situation to the virtual environment]. Bologna, Italy: Il Mulino.

Mantovani, G., & Riva, G. (1999). "Real" presence: How different ontologies generate different criteria for presence, telepresence, and virtual presence. *Presence, 8,* 540–550.

Mead, G. H. (1934). *Mind, self end society. From the standpoint of a behaviorist.* Chicago: University of Chicago Press.

Morris, D. (1977). *Manwatching.* London: Jonathan Cape.

Paccagnella, L. (2000). *La comunicazione al computer* [Communication with computers]. Bologna, Italy: Il Mulino.

Resnik, S. (1986). *L'esperienza psicotica* [The psychotic experience]. Turin, Italy: Boringhieri.

Ryff, C. D., & Singer, B. (2000). Interpersonal flourishing: A positive health agenda for the New Millennium. *Personality and Social Psychology Review, 4,* 30–44.

Scheflen, A. E. (1964). The significance of posture in communicative systems. *Psychiatry, 27,* 316–331.

Short, J. A., Williams, E., & Christie, B. (1976). *The social psychology of telecommunication.* Chichester, England: Wiley.

Slater, M., & Steed, A. (2000). A virtual presence counter. *Presence, 9,* 413–434.

Spencer, J. (2000). *Live arts experiences.* Retrieved January 27, 2002, from http://www.hospitalaudiences.org

Steuer, J. (1995). Defining virtual reality: Dimensions determining telepresence. In F. Biocca (Ed.), *Communication in the age of virtual reality* (pp. 73–93). Hillsdale, NJ: Lawrence Erlbaum Associates.

Uchino, B. N., Cacioppo, J. T., Kiecolt-Glaser, J. K. (1996). The relationship between social support and physiological processes: A review with emphasis on underlying mechanisms and implications for health. *Psychological Bulletin, 119,* 488–531.

Uvmas-Moderg, K. (1998). Oxytocin may mediate the benefits of positive social interactions and emotions. *Psychoneuroendocrinology, 23,* 819–835.

Wallace, P. (1999). *The psychology of the Internet.* Cambridge, England: Cambridge University Press.

CHAPTER FOURTEEN

The Next Frontier of Technology: Awaiting UMTS

Giovanni Strocchi

The Universal Mobile Telecommunications System (UMTS) is the third-generation mobile telecommunications standard. It will enable the full integration of mobile telecommunications and computing for the transmission of textual information and audio-video contents directly onto the display of mobile phones.

UMTS represents the final step in a process of technological development (see Table 14.1) that has progressively introduced an increasingly efficient data transmission capacity for voice services offered using the Global System for Mobile Communications (GSM) standard. In the near future this technology will offer the opportunity to integrate a range of services from simple telephony to the Internet, videoconferencing, and the distribution of audio-video contents on the same portable device—whose dimensions and appearance will presumably be more similar to those of today's palmtop computer.

The aim is to develop services with a high added value to offer to the large number of customers who already use mobile phones for voice communications (mobile telephony is one of the fastest-growing industries in the world). The competition in the provision of UMTS services will be above all in the sphere of innovative and creative ideas and the relative capacity to make these services available in ever more quickly and with excellent quality.

In Italy the race toward UMTS officially commenced on August 24, 2000, the expiry date for the presentation of applications to participate in the bidding for the assignment of the licenses, and ended in January 2001 with the definitive assignment of licenses to the successful bidders. The latter were granted a period of 30 months in which to activate the services in the regional capitals and a further 30 months in which to extend operations to the provincial capitals. The first UMTS services and new terminals became available to customers beginning in 2002. The current second-generation GSM and GPRS (General Packet Radio Service) net-

TABLE 14.1

The Future: Multimedia

	GSM	GPRS		UMTS	
	Now	Generation I (2000)	Generation II (2001)	Generation I (2002)	Generation II (2003–2004)
Current speed	9.6 Kb/s	52 Kb/s	~110 Kb/s	≥ 144 Kb/s	≥ 384 Kb/s
Increase in voice capacity	—	—	—	Y	Y
"Always on" data connection	—	Y	Y	Y	Y
Simultaneous voice/data	—	—	Y	Y	Y
"Real" broadcasting	—	—	—	Y (2000 release)	Y
Localization	Y	Y	Y	Y	Y
Security	Y	Y	Y	Y	Y

work, although with reduced quality and potential, will guarantee continuity of service with the previous standards.

1. THE TECHNOLOGICAL PLATFORM

UMTS is a technological platform that was born out of the idea of the International Mobile Telecommunication single world standard (IMT-2000), with the intention of providing a technology capable of making the interoperation of communications systems completely seamless. Only later was it decided to take a different course: The various bodies responsible for the definition of the standards developed proprietary systems, but with the obligation to make them interact perfectly with each other.

The maximum initial transmission speed (in reception) will be 40 times higher than the current digital GSM standard, 7 times higher than a normal domestic telephone line, and 3 times higher than a fixed ISDN connection.

The stages of development of the UMTS technology provide for the complete use of the Internet protocol (IP architecture) for the simultaneous control of both voice and data transmission over a single network. Despite the fact that the standard supports circuit-switching communication technologies—in order to make it compatible with the current standards used—its architectural platform has been defined in a manner strongly oriented toward packet switching and the Internet protocol. Communication via "packets" rather than "circuits" has various advantages. It constitutes a permanent logical connection that uses the transmission channels only when the information is being sent, leaving the terminal ready for communication at all times (known as "always on"). Furthermore, this connection enables an asymmetric transmission mode (in other words, it is possible to download information more rapidly than it is sent) aimed at aiding the reception of tones, images, and other data by the customer.

Specific contents and services will be created in order to exploit the new third-generation technological platform to the full, and the terminals will be substantially different from those currently in use, with larger, colored displays and user-friendly interfaces. The reproduction of multimedia contents will be optimized, with high-resolution images and video and high-fidelity stereo audio files. Tiny video cameras will be built into these new mobile phones, which will allow the filming of events and their successive transmission via email or videoconferencing. It will also be possible to equip the third-generation super mobile phones to provide navigation systems similar to those of the current satellite-guided global positioning system (GPS)—with which they will also be able to be integrated—and other locating-based services.

The analysts claim that with the arrival of UMTS technologies we will witness a boom in so-called m-commerce, that is, the "mobile" version of the electronic commerce familiar to us today. The pushbutton panel of the telephone will be transformed into a terminal for commercial transactions that will enable control over all the stages in the on-line purchasing process, from the search for the product to payment.

2. THE MOBILE PHONE AND CHANGES IN ITS PERCEPTION

The value attributed to the mobile phone and its perception by consumers have undergone great changes since 1991, the year in which the first mobile telephones appeared on the market. Initially the mobile phone was seen only as a professional tool, which was innovative but free of any symbolic connotations. The immediate expansion of the consumer market transformed this tool into a status symbol. With the introduction of the prepaid telephone card and the consequent opening up of the market to lower-income customers, the mobile phone acquired a social and publicly useful role. Indeed, it offered greater freedom of movement and action while at the same time permitting a greater degree of security and autonomy to those who use it. Finally, the availability on the market of extremely competitively priced handsets has enabled the widespread distribution of the mobile telephone, making it a mass phenomenon more than 41 million customers at the end of 2000. It has been transformed in the collective imagination from a "tool" to a "service" capable of aiding social and work relations.

Today the mobile phone cannot escape from becoming a symbolic representation of its owner. Fashions are thus created in the aesthetic quest for the model that reflects the image of its owner to the outside world. The mobile phone becomes a true symbolic accessory that few can do without, necessary for affirming one's uniqueness.

The introduction of the third-generation UMTS will result in another conceptual leap. The multiplicity of uses that will be possible in the future with this new tool will change the perception of the mobile phone itself in our collective imagination. The new models, equipped with larger displays in color and capable of transmitting images and video, will presumably take very different forms from those with which we are familiar, until they lose their mobile phone form and become multimedia terminals. Once they have lost the last image that conceptually linked them to their original function, it is obvious that their value will also become more abstract and further removed from the function for which they were created.

3. WHY UMTS?

Data transmission and mobile multimedia have become demands that operators in the field of mobile telephony cannot ignore. The market requires means of mobile communication that support increasingly sophisticated services and products capable of exchanging information and data with the outside world and accompanying and assisting the life of users during every moment of their mobile day.

Whereas today we are accompanied by numerous tools, each with a single function (e.g., identity card, diary, keys, and wallet), tomorrow it will be possible to integrate them all into a single multimedia terminal. Through this terminal we will be able to communicate, store information, plan our days, control our homes, cars, alarms, and more besides from a distance, or make banking transactions or purchases. This terminal will be our identity document, but it will also be the fastest means of surfing the Internet or playing games. The everyday quality of life should certainly improve considerably as a result of the opportunity to manage one's time, including working time, in an autonomous and personal manner, even while on the move.

With UMTS it will be possible to personalize mobile communications and the types of personal interaction will be multiplied, for example by sending images with text and music attachments. A new way of communicating will definitely be created whose developments and forms are still to be discovered.

PART III

Dressing Technologies

CHAPTER FIFTEEN

Slaves and Free People in the Gaily Colored Empire

Giorgio Pacifici and Paolo Girardi

An intimate connection between the body, fashion, and technology seems to occur not only in the individual sphere but also on the broader plane, where groups and microgroups are newly defining themselves. Developing our research on the impact of technological developments on both the individual psychological world and the social environment, we discuss the possible interaction of fashion and psychological problems, as we did in our previous research on technological stress[1] and on the world of hackers.[2]

Such interdisciplinary research leads to disputes about the analysis and interpretation of data. The attempt to solve this conflict by bringing in both "specialists" and "non-specialists" is a true methodological innovation.

Fashion, as a representative metaphor of the deepest elements of the being, is discussed now not only as a signal of sociological trends but also as a language that explains the adaptive and transformative modalities of human beings, and so is decipherable with the tools of the "psi" sciences (psychology, psychoanalysis, psychotherapy, psychiatry).

1. LOOK AT ME! LET ME SEE YOU!

In the post-Freudian sense, we can class as exhibitionism any behavior motivated by the pleasure of being seen and judged. Especially nowadays, we are inclined to consider exhibitionism in the broadest meaning, which includes hearing ourselves speak and displaying our possessions, as a frenzied defense against depres-

---

[1]International Meeting on "Work, technology and transformation of alienation," Treviso, September 20–22, 1990.

[2]Interdisciplinary research group, directed by Paolo Girardi and Giorgio Pacifici, on "Psychiatric, psychological, and social aspects of hacker group activities in Italy." University of Rome La Sapienza, Forum per la Tecnologia dell'Informazione (FTI), Associazione per la Ricerca Comparata e Interdisciplinare (ARCO).

sion, frigidity, impotence, or the fear of losing our own identity—in the sense of, "I am seen, therefore I am" (Fenichel, 1945).

Voyeurism is the pleasure that derives from seeing, and it mirrors the pleasure of being seen. Such pleasure, when sublimated, can be translated into intellectual curiosity, that is, into the thirst for knowledge; it is a pleasure that can manifest itself in an investigative attitude toward the environment or toward our inner world (De Martis, 1989).

In the exhibitionist and voyeuristic components of the human mind, so linked to sex, to the body, to the extremes of being dressed/undressed, and to the curiosity of acknowledgment, can be found the foundations of fashion. Spanish film director Almodovar's exhortation "Look at me!"—that is, "Look at my body, look at my beauty, look at my nakedness," but also "Look at the colors I wear, at my jewelry, at the skins on my skin"—expresses the essence of fashion, as stylists and fashion designers know well. However, at least a part of fashion's essence can also be found in saying "Let me see you!"—that is, "Let me see your beauty, your ornaments, the colors, the silks, the brocades, the sheer fabrics, and your naked body."

At least, exhibitionism and voyeurism *were* the foundations of fashion until the "reappearance" in contemporary society of a certain sadomasochistic component: a two-faced "perversion" whose active and passive form can usually be found in the same individual (Abraham, 1908/1975). The person who enjoys inflicting pain on others (in sex and, indeed, in all interpersonal relations) is also capable of enjoying the pain that results from these actions.

This tendency toward sadomasochism also shows itself in everything connected with the adjective *extreme*: extreme sex, extreme sports, but also extreme fashion, tattoos, piercing, branding. Think of the fashion for worn-out and dirty clothes, for false bloodstains, for slashes in clothes, for stains, for ostentatious patches, the fashion for pain/pleasure for the person wearing them, to be sneered at and to subject oneself to the stares of others. It is very easy to interpret this "grunge" aspect as postatomic mourning (Vetrugno, 1999); do not the Holy Scriptures order mourners "to tear their clothes" and "to abstain from washing"? And cannot all this black be interpreted as self-punishment and a mourning of the self? But in modern society there is even the possibility of this sadomasochist component becoming prevalent. What has spread is a feeling of uncertainty, of danger, which cannot be overcome.

2. HYPERCONSUMERS AND THE LUMPENPROLETARIAT OF FASHION

Fashion, then, imposes a harsh dividing line between two hostile and antagonistic populations, separated by faith but also by the economy.

The first group is composed of superior or "trendy" people, the believers. They follow fashion, frequent the "correct" restaurants, eat the "correct" foods, wear the "correct" clothes, and are therefore admired. They have an audience, a following—whether big or small. These followers, as in all Mysteries, are not yet admitted to the holy inner sanctum, but they may be in the future.

The second group are the unbelievers, the lumpenproletariat of fashion. They do not spend money on fashion (perhaps because they do not want to; perhaps because they do not have the money to), so they cannot be observed and admired, photographed, talked about, interviewed. They may or may not be "drop outs."

Unfortunately for the enthusiasts of the "trendy," it is possible for their position to be reversed by people who use different criteria. Some people—in line with the theories of the real genius of fashion, Franco Moschino (Moschino & Castelli, 1996)—create their own fashion, imaginatively and intuitively matching shapes and colors, raiding the past, utilizing, adapting, transforming. Such people, who in the past were perhaps slaves, are now free or at least freed people, and they mockingly and knowingly observe the slaves of modern fashion, with its corybantic designers and auletic media types.[3]

These modern-day slaves are by definition "fashion victims," and can be both women and men. They may be women compelled by misogynistic designers to augment or reduce their breasts; girls constrained to reveal their knobbly knees or flabby thighs; men forced into clothes that are too tight across the shoulders, or transformed into monsters by baggy trousers or into false Scotsmen by improbable kilts. All these things prevent them from looking comfortable and at ease.

But what is new in all of this? A lot and, at the same time, not a lot. On the one hand, it is sufficient to read the late 19th-century satirical magazines and to look at the pictures to realize how much the victims of fashion were also the victims of the humorists, who cruelly mocked their lack of common sense, or at least what common people considered common sense. On the other hand, fashion now resonates through technological communications media—TV, the Internet, chat lines, and newsgroups—to alert everyone to a sense of what is "possible." Because of this, fashion has acquired a depth and meaning that not even fashion historians could have predicted.

3. DRESS AND DISGUISE

Theoreticians say that "perversions" can be divided into "anomalies in the choice of the object" and "anomalies in the aim." In both categories we can make a further distinction, depending on whether that relation is established with a human being or with a symbolic substitute.

As regards "perversion with an anomalous aim," we can distinguish the axes sadism/masochism and exhibitionism/voyeurism (which we have already discussed) and fetishism/transvestitism, for which the participation of a partner is not required (Fenichel, 1945).

There probably is no popular magazine that has not dedicated a cover story to the world of transvestitism and to the theme of "dress and disguise." The trans-

---

[3]Corybants were priests of the goddess Cybele. A prominent feature of their ritual was a wild dance, which was claimed to have the power to heal mental disorder. "Auletes" in Greek antiquity were players of woodwind instruments (*aulos*).

vestite, with his wig and his glamorous dresses, has become a popular character on TV shows and at discos. He is not a damned soul but a "normal next-door neighbor," to be welcomed in leisurely evenings at home in front of the TV set. Today nobody whispers about the feather-trimmed dressing gown that the high-powered manager or civil servant wears in private, because the "perversion" has been assimilated and metabolized.

We may mention in passing the excessive femininity of transvestite fashion, which recalls the brilliance of the world of childish enchantment with its symbolic links and its attracting and repelling qualities. At the same time, transvestitism, which is also latent in some women's styles that make them look like plainclothes officers, is psychologically moderated by details and accessories that underline its amusing characteristics.

4. SHOPPING WITHOUT DESIRE

Compulsion is an irrational and coercive tendency to do things that individuals themselves recognize as useless and inadequate, but that, if not carried out, provoke anxiety. The symptoms of compulsion are characteristic of obsession neurosis and include any thoughts that individuals cannot suppress or behaviors that they are forced to accomplish.

Compulsions, like all symptoms, represent a compromise between instinctive demands and defensive forces. Instinct expresses itself in the intensity and immediateness of the demand. When instinct prevails, the impulse persists but it loses its value of desire and becomes a compulsive need.

The imperative to purchase objects that not only are unnecessary but often do not satisfy aesthetic ambitions either is a widespread phenomenon. The temples of compulsive buying are the department stores at "sales" time. Hundreds of documentaries have depicted the orgiastic rites of bacchantes inflamed by the necessity of buying. Gentle housewives, timid female students, and thoughtful clerks lose their wits and snatch dresses from one another's hands.

In the same way, a large part of the ritualized and unenthusiastic shopping in the salons of haute couture and ready-to-wear can be classified as compulsive buying. It is just that the fashion, the clothes, and the accessories are a more interesting theater of action for this syndrome.

Furthermore, any variation in color, in length, or in material of any article of clothing seems to offer an excuse—if a thin one—for another purchase. We might observe that in these cases, along with a compulsion to buy, there is a "compulsion to repeat," which urges "the slave" to reproduce behaviors, experiences, and situations already lived and to some extent already assimilated.

5. THE "INTRANEOUS" BODY

The expression "extraneous body" seems to be without content and meaning. The body is more and more involved in technological systems (e.g., virtual reality), just as technological devices increasingly intrude in the body (internal and external prostheses, small and large, shell and shelter). Prostheses such as eyeglasses or false eyelashes or nails all assert themselves as fashion. But any extension of the body—a pen-telephone, a watch, a personal computer- a car driven by voice commands—can become fashion. And fashion embraces everything, from silicon to plump up body parts to utilization of new materials to repair limbs.

The ability to use technology merges with the enjoyment of seeing and being seen. But at the same time the body is "intraneous" because it is becoming an intranet; the central and peripheral nervous systems work side by side with a net whose sensors and files are clothes and artificial prostheses. From the latest research at MIT's Medialab, we can observe that the human body is tending to become the privileged focus of technological renewal. One of the most interesting research areas is the Personal Area Network (PAN), which analyzes relations between the corporeal individual and information systems in a new way (Brand, 1987). In the PAN system, the body is the medium through which bits are transmitted, at a maximum speed of 100,000 bits a second, so our eyeglasses, for example, or our shoes can communicate. Our clothes become a net: our body/our clothing/our intranet (Zimmerman, 1996).

What are the latest developments in the field of wearable computers, and what repercussions can we expect at the mass-production level? Will intelligent clothes bring greater freedom or intense and acute informational stress?

Technologists at a multinational design company, the Lunar Corporation, are hard at work creating a jacket that could also function as the monitor of a PC. The central problem is creating a fabric that is able to translate the impulses into images without losing its lightness.

But other areas of research are equally important, such as applying the properties of computers to textiles. For example, scientists are studying how to obtain power from the heat that radiates from body movement, in order to store and provide information. Shoes that teach us how to dance, T-shirts that act as tourist guides, gloves that allow us to feel the substance of imaginary objects (as virtual reality gloves already do), will come on the market. This transformation has some consequences for the world of the fashion. All the research on the materials of the clothing and textile industry—on colors, on softness, on shapes—which until now have had almost exclusively aesthetic characteristics, must adapt itself to the technological movement.

The utilization of a certain fabric, the shape of a heel, or a particular hairstyle will no longer be just an expression of the personality of the stylist, symptoms of a mania for detail. Designers will have to take into consideration the characteristics of certain semiconductors, the possibility of accommodating a chip, the capability of being part of an intranet. According to the Italian architect and designer Stefano Marzano, by the year 2025 we will buy clothing not for its pretty color or its nice cut, but essentially for the digital equipment it includes. But this prediction could be overtaken by events.

6. THE UGLY JEWEL

As everybody knows, fashion is a form of social control. We may define social control as the totality of mechanisms that a community utilizes to prevent the "deviance" of an individual from a certain behavior, or to eliminate deviance that has already occurred; in other words, to ensure that individuals comply with social standards and rules (Merton, 1962).

It is a truism that the social group that holds power will make sure that citizens comply with social rules. Surveillance by electronic bracelet seems to be a perfect model for the merging of techno-clothing and conformity to legal strictures. But the notion of exercising control as a game, whether a game of love or a game of domination, could transform the e-bracelet into an item for the mass market. It could become a new extreme fashion, a new sort of "branding" or "piercing" marked by the sadomasochistic axe.

On the other hand, the bracelet and the anklet have an ancient use as symbols of domination and subjugation, with a strong sexual component. The bracelet is the final ring in a great chain of an imaginary and desired imprisonment ("Tie me!"). Bracelets and "slave sandals" have long been fashion objects. Today, love-slaves could make the e-bracelet a cult object, a symbol of their condition as postmodern slaves—always available and findable, even more so than through the mobile phone. And so the e-bracelet, once the subject of argument in the fashion media, could become a fashionable object, embellished by engravings and precious stones, and gain a share of the market.

7. HI-TECH: TOWARD A NEW AESTHETIC

In this short chapter we first interpreted the phenomenon of fashion socio-analytically, with a light post-Marxian reading. We looked at the mysterious structur of fashion, and its deep unawareness of a new and bloodless class struggle between "haves" and "have-nots." We only sketched the perspectives that information and communication technology give to the body and to fashion, making them one unique thing.

Will high-tech clothes permit us to arrive at a new meaning of beauty; to sublimate instinctive compulsions and to compensate for "perversions"?

An Italian journalist, Vittorio Zambardino, perceptively asked one of the new style gurus, Mr. Negroponte, during an interview if "wearable computers" would limit themselves to offering users broader access to technology, to giving them information, or if we can speak of the birth of a new aesthetics (Zambardino, 2000). The triumphant reply was that computers will both improve our quality of life and contribute to the creation of a new aesthetics.

But this new aesthetic will not be linked only to appearance and sensations. If people wear moveable objects, created by digital technology, this must also be a personality statement. If clothes communicate and assume the functions of a computer, it will be hard to keep knowledge, ethics, and fashion behavior separate.

Over the centuries, clothes have gradually developed from being body coverings, and no more, into fashion, a means to project an image. Clothes that communicate seem to presage a new and different phase. It is possible that this technological evolution will be guided by the tyrannical "hand of the market." But it also possible that it will be dominated by "philosopher kings," experts in bionics, material sciences, and marketing laws, but above all devoted to serving the evolution, psychological health, and social evolution of human beings.

REFERENCES AND BIBLIOGRAPHY

Abraham, K. (1975). *Differenze psicosessuali tra isterici e dementia precox* [Psycho-sexual differences in hysteria and dementia precox.]. In *Opere* [Works] (pp. 93–108), Turin, Italy: Boringhieri. (Original work published 1908)

Abraham, K. (1975). *Opere* [Works] (Italian translation). Turin, Italy: Boringhieri.

Andrieu, B. (1993). *Le corps dispersé. Une histoire du corps au XXᵉ siècle* [The dispersed body. A history of the body in the 20th century]. Paris: L'Harmattan.

Bataille, G. (1949). *La parte maledetta* [The cursed part]. Verona, Italy: Bertani.

Bernasconi, S. (1974). Marucelli. In P. Soli, *Il genio antipatico. Creatività e tecnologia della moda italiana 1951/83* [The unpleasant genius. Creativity and technology in Italian fashion 1951–83] (pp. 43–61). Milan: Catalogo della mostra.

Borel, F. (1992). *Le vêtement incarné. Les métamorphoses du corps* [The incarnated garment. Metamorphoses of the body]. Paris: Calmann-Lévy.

Bovone, L., & Mora, E. (1997). *La moda della metropoli* [Fashion in the metropolis]. Milan: Angeli.

Brand, S. (1987). *The Media Lab. Inventing the future at MIT.* New York: Viking.

Brohm, J. M., Larrère, C., & Lascoumes, P. (1992). *Corps. Sociétés, sciences, politiques, imaginaires* [Body, societies, sciences, politics of the imagination]. Paris: Belin.

Butler, J. (1993). *Bodies that matter. On the discursive limits of "sex."* London: Routledge.

Capucci, P. L. (Ed.). (1994). *Il corpo tecnologico. L'infmuenza delle tecnologie sul corpo e sulle sue facoltà* [The technological body. The influence of technologies on the body and its faculties]. Bologna, Italy: Baskerville.

CENSIS (1994). *Moda & comunicazione. I protagonisti e le prospettive di un sistema che cambia* [Fashion and communication. Protagonists and perspectives in a changing system]. Milan: Angeli.

Ceriani, G. (Ed.). (1995). *Moda: regole e rappresentazioni* [Fashion: Rules and representations]. Milan: Angeli.

Curcio, A. M., & Ferrarotti, F. (1994). *La moda: identità negata* [Fashion: Denied identity]. Milan: Angeli.

De Kerckhove, D. (1994). Remapping sensoriale nella realtà virtuale e nelle altre tecnologie ciberattive [Sensorial remapping in virtual reality and other cyber-active technologies]. In P. L. Capucci, (Ed.), *Il corpo tecnologico. L'influenza delle tecnologie sul corpo e sulle sue facoltà* [The technological body. The influence of technologies on the body and its faculties] (pp. 73–129). Bologna, Italy: Baskerville.

De Martis, D. (1989). La perversione, A) Aspetti generali [Perversion, general aspects]. In A. A. Semi (Ed.), *Trattato di psicanalisi* [Treatise of psychoanalysis] (Vol. 2, pp. 263–265). Milan: Raffaello cortina Editore.

Durkheim, E. (1971). *Le forme elementari della vita religiosa* [Elementary forms of the religious life] (Italian translation). Milan: Comunità. (Original work published 1912)

Ewen, S. (1993). *Sotto l'immagine niente. La politica dello stile nella società contemporanea* [Under the image nothing. The politics of style in contemporary society]. Milan: Angeli.

Featherstone, M., & Borrows, R. (1995). *Cyberspace/cyberbodies/cyberpunk. Cultures of technological embodiment.* London: Sage.

Fenichel, O. (1945). *The psychoanalytic theory of neurosis.* New York: Norton. (Italian translation, *Trattato di psicoanalisi delle nevrosi e delle psicosi.* Rome: Astrolabio, 1951.)

Flugel, J. C., & Tibaldi, G. (1992). *Psicologia dell'abbigliamento* [Psychology of clothing]. Milan: Angeli.

Frazer, J. G. (1973). *Il ramo d'oro (1911–1915)* [The golden bough (1911/1915)]. Turin, Italy: Boringhieri.

Galimberti, U. (1987). *Il corpo* [The body]. Milan: Feltrinelli.

Gastel, M. (1995). *50 Anni di moda italiana* [50 years of Italian fashion]. Milan: Vallardi.

Harris, T. A. (1976). *Io sono ok, tu sei ok* [I'm OK, you're OK]. Milan: Rizzoli.

Le Breton, D. (1990). *Anthropologie du corps et modernité* [Anthropology of the body and modernity]. Paris: Presses Universitaires de France.

Le Breton, D. (1991). *Corps et sociétés* [Bodies and societies]. Paris: Méridiens Klincksieck.

Le Breton, D. (1992). *La sociologie du corps* [The sociology of the body]. Paris: Presses Universitaires de France.

Maldonado, T. (1997). *Critica della ragione informatica* [Critique of IT reason]. Milan: Feltrinelli.

Mangili, F., & Musso, G. (1992). *La sensorialità delle macchine* [The sensorialness of machines]. Milan: McGraw-Hill.

Mehrabian, A. (1972). *Nonverbal communication.* Chicago: Aldine Atherton.

Merton, R. K. (1949). *Social theory and social structure.* New York: The Free Press. (Italian translation, *Teoria e struttura sociale.*) Bologna: Il Mulino, 1983.

Moschino, F., & Castelli, L. (1996). *X anni di caos* [Ten years of chaos] (3rd ed.). Milan: Edizioni Lybra Immagine.

Pacifici, G. (1999). *La moda italiana tra grande business internazionale e fenomeno culturale di massa* [Italian fashion: International big business and the phenomenon of mass culture] (Corso svolto nell'anno accademico 1998–1999 [Course held in the academic year 1998–1999]). Paris: Institut d'Études Politiques.

Passerini, M. (1994). *Nuova enciclopedia della moda italiana* [New encyclopedia of Italian fashion]. Rome: Eidos Edizioni.

Perrot, P. (1984). *Le travail des apparences. Ou les transformations du corps féminin au XVIIIe–XXe siècle* [The work of appearances. Or the transformations of the female body in the 18th–20th centuries]. Paris: Seuil.

Ragone, G., Alberoni, F., & Barber, B. (1992). *Sociologia dei fenomeni di moda* [Sociology of phenomena of fashion]. Milan: Angeli.

Simon, D. (1990). *Moda e sociologia* [Fashion and sociology]. Milan: Angeli.

Velter, A., & Lamothe, M. J. (1978). *Les outils du corps* [The tools of the body]. Paris: Denoël/Gonthier.

Vergani, G. (Ed.). (1999). *Dizionario della moda* [Dictionary of fashion]. Milan: Baldini & Castoldi.

Versace, G. (1997). *The art of being you.* Milan: Leonardo Arte.

Vetrugno, M. (1999). Grunge (voce). In G. Vergani (Ed.), *Dizionario della moda* (pp. 43–58). Milan: Baldini & Castoldi.

Zambardino, V. (2000). Sarà e-commerce? [Can it be e-commerce?]. *Moda & Internet* [Fashion & Internet], Supplement to *Republica,* March.

Zimmerman, T. (1996). Personal area networks (PAN): Near-field intra-body communication. *IBM Systems Journal, 35,* 609–618.

CHAPTER SIXTEEN

Inside the Surface: Technology in Modern Textiles

Elda Danese

Textiles and technology have long been connected: Textile production was one of the first human tasks to be completely mechanized. Interestingly, technological progress in weaving was also instrumental in the development of the computer. In particular, the mechanical loom invented by the French weaver Joseph Marie Jacquard at the end of the 18th century, and the analysis of its punch cards on the mechanical weaving loom, inspired Charles Babbage in the 19th century to undertake the investigations that precipitated the information technology revolution (Babbage, 1835/1986).

Chemical research and production were also important to textiles. After the discovery of chemical dyes, such as William Henry Perkin's *mauveine* in 1856 or Emanuel Verguin's *fuchsine* in 1858 (Brunello, 1968), new artificial and synthetic fibers were invented. From the cellulose-based fiber patented by Chatillon in 1884 to those derived from petroleum, there has been a constant search for fibers that could substitute for natural ones (Fortunati, 1998). Synthetic fiber production has grown despite the fact that its aesthetic and functional characteristics have been seen as inferior to those of natural fibers. Indeed, by 1999, synthetics accounted for almost 60% of the entire production of textile fibers.

Accompanying this technological development has been scholarly research on the social and functional aspects of fabric. For instance, textile specialists Sarah Braddock and Marie O'Mahony (Braddock & O'Mahony, 1998) and Chloë Colchester (Colchester, 1991) have demonstrated to designers and scholars the innovative developments in the textile field, and helped to define the state of the technological and aesthetic evolution in textile production. The research carried out during the 1970s in Italy by the Centro Design Montefibre placed textile culture and decoration into a wider context, not exclusively related to clothing. A team of industrial designers edited a series of publications entitled *Decorattivo* (Centro Design Montefibre, 1975, 1976, 1977), which inspired innovations in the social culture of textiles.

Paralleling modern design endeavors, military and aerospace research has made many new materials commercially available, enabling innovation in textiles. The properties of this new generation of materials are especially useful in various industrial sectors such as building, transportation, and road building (Braddock & O'Mahony, 1998). These high-performance "technical textiles" constitute a rapidly growing portion of textile production. They are increasingly substituted for rigid, heavy materials such as metals. Superpolymers, inorganic fibers (based on carbon, glass, or ceramics), aramidic fibers, and metallic fibers are among these new high-performances textiles, whose uses are constantly being expanded.

An example of a technological by-product of aerospace research is triaxial fabric, a material developed in the late 1960s by the National Aeronautics and Space Administration (NASA) for pressure suits to be used aboard the Space Shuttle ("Space shuttle pressure suits," 1981). The intersection at 60 degrees of two warps and one weft yields not only greater stability and strength than the square right-angled intertwinement of fabrics, but greater flexibility as well. Such a structural design is traditionally used for hand-woven straw or wicker baskets. To make the continuous fabric surface necessary for space applications, a specially designed loom had to be created (McCarty & McQuaid, 1998). By utilizing composites of carbon fibers or fiberglass, the triaxial fabric is used for solar panels, skis, safety clothing, tension structures, and prosthetic devices for people with articulation problems (Manzini, 1989; Morley, 1987).

Another structural innovation, which is being studied for both its technical aspects and its expressive potentialities, is giving three-dimensionality to textile surfaces. The German firm Shape 3 has devised a weaving system that generates three-dimensional form while weaving the fabric. With this procedure, the cloth does not need to be subsequently treated or shaped, and in this way the surface is not put under stress; at the same time, some phases of the production process are abolished and material wastage is reduced (Braddock & O'Mahony, 1998).

"Technical textiles" are not confined to specialist applications, however. They often reach the public at large through sports equipment and clothing. A firm such as Gore, whose textile division is mainly known for Gore-Tex, a fabric widely used for leisure wear, also makes filters, parts of pumps, gaskets for pipes, gauze, medical materials, and many other specialist articles. Sometimes, the special properties of some fabrics are given a new function, which becomes the basis of a corporation's marketing. For instance, a fabric incorporating carbon particles, created in the 1970s with the aim of discharging static electricity from the work wear of Japanese workers, later entered the fashion circuit as an "anti-stress" material (Guillaume, 2000).

One reason for this shift in use in technologies is that the enterprises responsible encourage the application of new materials in different fields in order to gain a higher return on the large investments required for their development. In addition to the desire for novelty inspired and satisfied by the fashion machine, another explanation for this shifting of function in technologies is that the boundaries between living and working are becoming increasingly blurred; the demand for exchange of information between the sectors of sportswear, fashion, and other industries has grown parallel to the progressive overlapping of leisure time and working time.

Fashion designers and the clothing industry have shown great interest in technical textiles, based on both their functional characteristics and their appearance. According to the experts in the field, "today the stylistic research of fashion goes hand in hand with the research in textiles" (Bogo, 1999, p. 30). This opinion is confirmed by Giorgio Armani, who stated, "the new materials stimulate creation because they offer a great variety of interpretations and applications" (Bogo, 1999, p. 30).

To industrial sectors the technical performance of textiles is of the utmost importance, whereas in the case of clothing production the study of tactile and visual qualities is at least as important. Many European and American textile companies are further motivated to conceive of process and product innovations by the competition from Asian countries. Such tendencies are expressed in particular in the finishing phases, through the introduction of treatments and processes that give a new aspect and a new tactile quality to fabrics (Colchester, 1991). With these treatments, for instance, silk becomes "peach skin" or "angel skin," appellatives that relate to tactile sensations. Interestingly, this semantic exchange between body and fabric is reversed when one speaks of "velvet skin" or "silk skin."

The recent innovations in the field of textile research and production not only emphasize the introduction of technologies, but also institute the representation of technology in textiles or its visibility. Some of the materials that we wear nowadays, with regard to both texture and color, have a new consistency, a new surface treatment that does not imitate or evoke natural fibers. The technology is not hidden; it becomes an aesthetic option as garments exhibit a high technological content. Previously the brightness and metallic reflections of textiles came from gold and silver, and distinguished the richness and prestige of the people wearing them. Today, the iridescence and the brilliance of the coat spattered with titanium or of stainless steel–speckled organza—fabrics that protect against electromagnetic waves and infrared or ultraviolet rays, and that are soldered instead of sewn—reveal the seductive side of technology and represent a new form of luxury.

The fact that a technology can reveal itself through its contact with the skin, and that it can reveal a value, shows how technologies are becoming part of everyday life, albeit with social undertones. This new relationship with technology is channeled through a plethora of gadgets and toys, through user-friendly objects that not only talk but squeak and cry. The bright colors available for automobiles, telephones, and computers, and the rounded design of machines that brings into play comic nuances, are also ways of taming technology and making our relationship with these devices less threatening.

In spite of the widespread visibility of these advancements in textile technology, there are also counterexamples that allude to different points of view. The growing smoothness of materials is contrasted by creasing; the new and the indestructible are opposed by threadbare and tattered clothing. The technological euphoria of the durability and imperishability promised by carbon fibers and superpolymers is opposed by the image of the organic perishability of the oxidized and moldy fabric that the English designer Hussein Chalayan created by keeping it buried for a few months in a garden. The fashion industry, in producing new "used" clothes, is taking over the spontaneous trend of worn-out and torn jeans and the recycling of old dresses. The wear and tear signal the elapsing of time: in a positive sense it represents the experience and intensity of life; on the other hand, it can express a certain indifference toward the mutability of taste and outward appearances. The latter attitude has in turn been absorbed by the fashion system with respect to production and communication. For those who want to seem unconcerned by appearances, and for those who do not have time to wait for the passing of time, the fashion industry has devised processes such as stone-washing, which gives jeans a worn look in a few minutes, or "delavé," a technique for the rapid fading of fabric colors.

The need to see the elapsing of time through material objects, for instance in the case of secondhand clothes, is related to the concept of an ecological conscience. Contemporary fashion underscores this tendency to utilize and incorporate used clothes. For instance, the Belgian designer Martin Margiela has created a sweater from a patchwork of salvaged U.S. army socks. Old clothes have always been used to produce new ones. An historical example is represented by the reprocessing activity typical of the Italian textile district of Prato in the 19th century, where *cenci di lana* (wool rags) were unraveled to retrieve the fiber in order to produce new clothes. In that case, however, the regenerated nature of the textiles was not explicit, whereas recent fashion designs denote an emphatic and educated recycling. We can compare the recent creation of "perfectly spoiled" clothes— ripped fabrics and loose, holed, and laddered sweaters—to the term *sprezzatura*. This word, originating in the 16th century (Davanzo Poli & Moronato, 1994), suggested spontaneity and an effortless display of virtuosity; at the same time, however, it signified a nonchalant attitude toward social conventions and riches, expressed by the vogue for slashing the fabric of clothes.

The culture of uniformity, guaranteed perfection, and durability (which is contradicted, however, by the strong demand for change imposed by the rhythm of fashion) is in contrast to these images of perishability. In spite of the attempt to attribute value to imperfection (which in handicraft products is considered a manifestation of the embodied human work), industrial production has deemed characteristics of uniformity and duplicability synonymous with quality. The lack of irregularities and the homogeneity of the surface and color of a fabric are some of industry's important evaluation parameters.

These parameters, which are typical of Western culture, are, however, far from the reference point for Japanese textile designers, who, like Jun'ichi Arai (*Jun'ichi Arai and Nuno No Kodo*, 1997) and Reiko Sudo, combine the study of traditional techniques and the use of advanced technologies with the aim of obtaining unusual and unpredictable results. A particular characteristic of their activity is their interest in the plasticity of textiles and the expressive aspects of three-dimensionality. Among their experimental creations we find fabrics where protruding forms are heat-molded in synthetic materials to form a regular pattern; or where one of the different component fibers shrinks when treated with heat, resulting in strong visual and tactile contrasts.

The same designers, working with companies such as Nuno and Inoue Pleats (McCarty & McQuaid, 1998), have studied the possibility of modeling the surfaces of fabrics by pleating, creasing, and goffering them. These effects have an obvious decorative function at the design level, but at the same time could be seen as an interesting answer to the problem of clothing care, a reply to Sanforized (mechanically preshrunk material) and noniron fabrics.

This aspect was explicitly taken into consideration by Nanni Strada, who in 1986 created the "Torchon" collection of linen dresses, whose irregular pleating was maintained by twisting and knotting them inside a foulard when putting them away. Similar endeavors led her, in 1993, to design "Pli-Plà," a collection of dresses that can be folded like fans thanks to the seams that run along their entire length (Strada, 1998). In 1974 at the Milan Triennale, Strada presented her research on a production system for seamless polyamide-knitted garments (an adaptation of the production technology of pantyhose) (Strada & Castelli, 1974). Her research also addressed the subject of clothing with a progressive methodology, and stressed a fundamental aspect of contemporary clothing—its elasticity, which is the most important characteristic marking the boundary between natural and artificial fibers.

The pleats and creases that modulate the new Japanese fabrics, at the same time allowing them to be extended, are forms defined by the material. Its characteristics determine the shape of dress; clothes made with these fabrics are not conceived only for the model body. This principle relates to the concept of clothing in the traditional Japanese culture, the idea that clothes "have a life and a spirit of their own, interacting with the human body both in motion and in rest, and are versatile enough to adapt themselves to the lifestyle of the people wearing them" (Wada, 1995, p. 35). This concept is not common in Western culture where, according to Baudrillard (1979), "clothes must signify the body," a body that "is removed and signified allusively" (p. 118). This is all the more evident when our "second skin" stretches around our body, holding us and compressing and emphasizing any prominent design feature.

The new stress-resistant fibers are more elastic than our skin itself. A thread of elastomeric fiber can be extended up to six times its original length; tiny amounts of these yarns are imperceptibly included in a great number of fabrics in order to improve their performance, in particular their ability to maintain their shape. Of particular importance is the fact that the appearance of these materials reveals the technology embodied in them. Synthetic fibers, once extolled for imitating natural fibers, are interwoven in the new textiles that emphasize the shape and movement of the body. Moreover, their properties have been transformed by the introduction of microfibers, which give a greater softness to these and other fabrics. Microfibers are the outcome of the study of microstructures in nature. One way to make microfibers is by producing a bicomponent thread that is then halved in thickness by dissolving one of the materials composing it. These threads are 60 times finer than a human hair, and super-microfibers have a diameter of a thousandth of a micron.

Many researchers have focused their attention on textiles that are actually capable of holding and releasing substances and materials in contact with the body. Bactericides, cosmetics, and drugs are contained and released by microcapsules embedded among the fibers. In some cases, the tiny capsules break gradually during wear and release active ingredients that are absorbed by the skin. In other instances they have a rigid case, like those of Outlast,® one of the first fabrics capable of reacting to changes in physical and environmental conditions with a system devised for NASA by Triangle Research. The capsules contain paraffin, a phase change material (PCM), which absorbs the heat of the body and then releases it when the external temperature drops. In this fabric, as in other "intelligent textiles," the threads contain electronics, including PCMs, electrical conductors, and microchips. This is one route that has led to the recent creation of *wearable computers*—clothes and accessories integrating computers, mobile phones, and other technological devices that communicate by means of a *body area network*.

The performances required of new artificial textiles today resemble those granted by sensitive organisms such as human skin. Reciprocally, researchers working on the reconstruction of human skin utilize knowledge and technologies peculiar to textiles. Of crucial importance in the field of tissue engineering is the quest for biocompatible and biodegradable polymers. After concentrating on nondegradable synthetic polymers (PE and PET), the research focused first on degradable synthetic polymers (PGA and PLA) and then on natural polymers that are completely bio-interactive, biocompatible, and bio-absorbable. A new insoluble polymer used for producing threads, membranes, gauze, and sponges, for instance, is derived from the processing of hyaluronic acid. Up to now the nonwoven structure seems to be the form best suited to inducing cellular infiltration and growth, even if more complex structures are currently under investigation. Some of these materials, used for the reconstruction of cartilage and the growth of epidermal and dermal cells, are produced with looms specially adapted for the processing of biopolymers (Fidia Advanced Biopolymers, 1992; Miglino, 2000).

Some biotechnology companies, moreover, have formed joint ventures with large chemical enterprises to carry out research programs with the aim of improving or modifying the properties of natural fibers. In the case of "Bt-cotton," for instance, the genes of a bacterium that produce a toxin lethal to some species of insects are inserted in the plant; other researchers are trying to obtain colored cotton fibers directly on the plant through genetic manipulation. Genecor and DuPont have genetically modified and patented a new microorganism capable of transforming glucose into 1,3 propanediol (PDO), a polymer intermediate indispensable for manufacturing a new synthetic fiber. Nexia Biotechnologies, a Canadian company, is developing biosteel, a fiber derived from the milk of goats in which the production of spyrodroine is genetically stimulated. Along the same route, DuPont researchers are trying to make new protein-based polymers using recombinant DNA techniques. By inserting a gene of a web-weaving spider into silkworms, they hope to create an ultrafine and very strong silk (Guerette, Ginzinger, Weber, & Gosline, 1996). The interest and investments of many companies are focused on this frontier of biotechnology research, because they believe that it has the potential radically to transform production processes. The reason for this interest can be inferred from the following words of Ellen Kullman, vice president of DuPont Safety & Protection: "DNA is the quintessential production platform. It replicates itself the same way every time. When you use it to help produce something, it makes very few mistakes. This efficiency lends tremendous power over the traditional chemical route to polymers, which is a comparatively random process" (Kullman, 2000).

REFERENCES AND BIBLIOGRAPHY

Babbage, C. (1986).*On the economy of machinery and manufacture.* New York: Fairfield. (Original work published 1835)

Baudrillard, J. (1979). *Lo scambio simbolico e la morte* [Death and symbolic exchange]. Milan: Feltrinelli.

Bogo, D. (1999, September 24). Nella giacca c'è un sacco a pelo [The jacket hides a sleeping-sack]. *Corriere della Sera,* pp. 21–43.

Braddock, S. E., & O'Mahony, M. (1998). *Techno textiles. Revolutionary fabrics for fashion and design.* London: Thames & Hudson.

Brunello, F. (1968). *L'arte della tintoria nella storia dell'umanità* [The art of dye-works in the history of mankind]. Vicenza, Italy: Neri Pozza Editore.

Centro Design Montefibre. (1975). *Decorattivo 1975* [Decor-active 1975]. Vicenza, Italy: Montefibre S.p.a.

Centro Design Montefibre. (1976). *Decorattivo 1976* [Decor-active 1976]. Vicenza, Italy: Montefibre S.p.a.

Centro Design Montefibre. (1977). *Decorattivo 1977* [Decor-active 1977]. Vicenza, ,Italy: Montefibre S.p.a.

Colchester, C. (1991). *The new textiles. Trend + traditions.* London: Thames & Hudson.

Davanzo Poli, D., & Moronato, S. (1994). *Le stoffe dei Veneziani* [The materials of the Venetians]. Venice, Italy: Albrizzi Editore.

Fidia Advanced Biopolymers. (1992). *Novel biomaterials. Moving toward an industrial dimension.*

Fortunati, L. (1998). Revêtir des technologies [Wearing technologies]. *Reseaux, 90,* 28–40.

Guerette, P. A., Ginzinger, d. G., Weber, B. H. F., & Gosline, J. M. (1996). Silk properties determined by gland-specific expression of a spider fibrion gene family. *Science, 272,* 112–115.

Guillaume, V. (2000). La marche de la technologie, la poesie de la creation, 1960–2000 [The march of technology, the poetry of creation, 1960–2000]. *Mutations/mode 1960–2000.* Catalogue of the exhibition at the Musée de la Mode de la Ville de Paris, Paris.

Jun'ichi Arai and Nuno No Kodo. (1997). Catalogue of the exhibition at the Nederland Textielmuseum, Tilburg, Netherlands.

Kullman, E. (2000). Making the impossible possible. *DuPont magazine on line.* Retrieved from http://www.dupont.com/corp/products/dupontmag/00/making.html

McCarty, C., & McQuaid, M. (1998). *Structure and surface. Contemporary Japanese textiles* (Catalogue of the exhibition at the Modern Art Museum of New York).

Manzini, E. (1989). *La materia dell'invenzione* [Matter of invention]. Milan: Arcadia Edizioni.

Miglino, N. (2000). Tessuti artificiali made in Italy [Artificial fabrics made in Italy]. *Le Scienze dossier,* no. 4, Summer.

Morley, J. C. (1987). *High-performance fibre composites.* London: Academic Press.

Recycling: Forms for the next century. Austerity for posterity. (1997). Catalogue of the exhibition at the Craft Council Gallery, London.

Space shuttle pressure suits. (1981). *Spinoff,* January, p. 73.

Strada, N. (1998). *Moda design* [Fashion design]. Milan: Editoriale Modo.

Strada, N., & Castelli, C. T. (1974). Il manto e la pelle [The mantle and the skin]. *Casabella, 387,* 83–95.

Wada, Y. I. (1995). *Tessuti e kimono. Design giapponese, Una storia dal 1950* [Fabrics and the kimono. The story of Japanese design to 1950]. Catalogue of the exhibition at the Milan Triennale. Firenze, Italy: Octavo.

CHAPTER SEVENTEEN

Fashion, Media, and Cultural Anxiety: Visual Representations of Childhood

Susan B. Kaiser

More than 150 years ago, Kierkegaard (1980) described anxiety as the "ambiguity of subjectivity" (p. 197). He reasoned that anxiety corresponded to the future and to the "infinite possibility of being able" (p. 44). Abstractly, he argued, subjectivity lacks content; it is a moving "magic picture" (p. 159) in search of adventure. In Kierkegaard's view, Hegelian logic could not account for anxiety; this logic was unable to accommodate ambiguous transitions such as impatiently wanting to act and looking toward the future with a feeling "in one's bones that a storm is approaching" (p. 115). In Kierkegaard's terms, it is the dreaming spirit, not merely a dialectic, that anticipates the need to posit a synthesis, uniting the "temporal and the eternal" as well as the mind and body (p. 81). But this spirit also disturbs self-representations in and through time.

Kierkegaard's conceptualization of anxiety offers potential for a theoretical analysis of fashion. He described anxiety in a way that parallels contemporary theories highlighting themes of ambivalence and ambiguity (see, e.g., Boultwood & Jerrard, 2000; Davis, 1992; Kaiser, Nagasawa, & Hutton, 1991; Wilson, 1985). Fashion, I argue, becomes a vehicle through which contemporary cultures articulate the ambiguities of collective subjectivity in and through time. Via the media, fashion articulates cultural anxieties at the intersections of visual technologies, temporality, space/place, and self–other relations. Fashion anticipates and represents that which might otherwise go unspoken, including cultural anxieties that are intangible or freely floating. Ambiguous though they may be, fashion's expressions of cultural anxiety allude to deep contradictions in need of expression and attention.

Among fashion's expressions are the complex relations among the past, present, and future. The sociologist Herbert Blumer (1969) saw modern fashion as offering a means of detaching from the past, connecting with the present, and anticipating the future. Yet Kierkegaard (1980) problematized such temporal delineations, suggesting instead that "no moment is a present"; it is only when a culture spatializes time into segments that its infinite succession of moments comes to a halt (p. 85). The etymological relationship of "moment" to "momentum" suggests the ongoing movement of time and subjectivity. It is only through representation, Kierkegaard suggested, that the moment becomes posited as a present in relation to the categories of the past and the future. Today's visual technoculture (e.g., films, cable television, the Internet) saturates everyday life with representational moments—marking anxieties in the process.

For example, Hethorn and Kaiser (1998) discussed how "youth style" serves to articulate cultural anxieties; it does so in a way that both reflects and constitutes cultural uneasiness about adolescent sexuality and violence. The way adolescents dress in everyday life can be analyzed discursively in relation to music videos, fashion advertising, and other media. That is, young people articulate complex ideas visually through their appearance styles, using and adapting goods available in the marketplace. They also get ideas from technoculture. However, given their need to experiment with their identities, rarely do they adopt any look *exactly* as it has been presented to them. Rather, they innovate and negotiate "looks" and meanings in everyday interactions with one another. The apparel industry and technoculture appropriate some of these looks and meanings and adapt them for further consumption.

Adolescence represents a transitional space between childhood and adulthood, and at least some of the anxiety associated with this space can be attributed to the ambiguous nature of its "in between" status. The emerging "'tween" clothing market—approximately targeted at girls aged 7 to 12 and boys aged 8 to 14—is further truncating the years of childhood and is contributing to the growing concern that children are growing up too fast. (The 'tween category is one of the fastest-growing consumer markets, estimated to influence more than U.S.$150 billion in annual U.S. family spending—Azolay, 2001.)

In the remainder of this chapter, I use the case study of fashionable childhood to revisit Kierkegaard's (1980) premise that anxiety is found "in all cultures where the childlike is preserved as the dreaming of the spirit" (p. 42). Anxieties about the preservation of childhood abound in U.S. cultural discourse, which suggests that the "age of protection" has given way to an "age of preparation" (Winn, 1981). Conservative cultural critics such as Medved and Medved (1998) caution that children are being exposed to "every perversion and peril in an effort to 'prepare' them for a harsh, dangerous future" (p. 1). (Concerns about preserving childhood are not limited to the far right, however. Former U.S. Vice President Al Gore's wife, Tipper Gore, for example, has spoken publicly and written about the perils of raising children in an "X-rated society.") They cited the premature death of the child beauty queen JonBenet Ramsey in Boulder, Colorado, as an example of the culture of preparation and a "paradoxical fascination in our society with precocity" (p. 7). They argued that her seductive, made-up image and "posed pout, splashed like a coldly calculated fashion model's smirk across so many magazine covers, both captivated and disturbed us" because of the paradoxical "pairings of innocence and seduction, childhood and death" (p. 9). The Medveds further directly implicated fashion in the stimulation of an impatience for children—boys as well as girls—to grow up too fast. They nostalgically recalled the 1970s:

Remember the happy face, popular in the 1970s? Those were the days when colorful Hawaiian-print surf shirts ruled clothes racks, when that unsophisticated circle with its line-smile was everywhere, along with the salutation: "Have a nice day!" By contrast, what slogans dominate now: No Fear. Scowling Bad Boys. "Gangsta" clothing that hangs so baggy off kids' hips that they can barely waddle. This is the climate that has been fostered by rap music and its drone of complaint and rage. We were stunned to see a fashion-conscious suburban mother dress her two-year-old in a miniature "gangsta" outfit, with wide baggy pants and a stylishly torn T-shirt. The toddler's hair was cut and greased so that it shot up in short spikes from his pudgy preschool face. No, dressing up a child isn't major, but where does a two-year-old punk go from there? (p. 78)

The Medveds (1998) argued that the attitude and "edge" in the appearances of young children are "at odds with the childhood we want for our children.... [T]he only unencumbered time they'll have to invent their own images, to explore their own fantasies, to create their own possibilities, is the few precious years before puberty" (p. 94).

I would argue that this is exactly what the inner-city, minority males who initially invented hip-hop style were doing: exploring their own fantasies and creating their own possibilities. In doing so, they developed ways of representing themselves that both resist and accommodate dominant ways of being. What the Medveds seem to miss in their analysis is the agency of young people—'tweens and teens alike—in the realm of appearance style.

It is perhaps telling that much of the anxious cultural discourse about representations of childhood revolves around (a) styles associated in the public imagination with minority males and violence and (b) images of young White, blond girls made up to look seductively adultlike. Discourses of race, class, and sexuality, as well as gender, intersect in complex ways that not only frame cultural anxieties surrounding youth style, but also render the *content* to subjectivity that Kierkegaard described as otherwise lacking. I would submit that cultural representations of childhood serve to articulate the ambiguities of collective subjectivity—anxieties about the future that style and fashion are able to capture in what Kierkegaard described as a moving, magic picture that uneasily posits the moment as a present in relation to the future.

In representations of infancy, the positing of the cultural moment often involves an uncomplicated pairing of gender and innocence. As Spencer (1995) noted in his history of homosexuality, "Our children are sacrosanct. We do not think of them as sexual beings, yet we do think of them very distinctly as boys and girls. Their gender is important to us. We believe strongly that all children are innocent and must be kept that way for as long as possible" (p. 404). Infants were used in all their gendered innocence in a Macy's newspaper ad on January 1, 2000, to depict the new century's "world of open possibilities." Everything was open, presumably, except their genders. Pink and blue banners, and a tiara for the girl, adorn the diapered babies; they attend to one another with innocent interest.

By the 'tween years—the apparel industry's new age category for sizes 8 to 14—if not sooner, gender can no longer be easily separated from sexuality. Despite protestations by some representatives of the apparel industry that 'tween girls are supposed to look like teenage girls "without being provocative" (Azolay, 2001, p. 47), cultural anxieties revolve around premature sexualization. In a clothing ad in the industry periodical *Children's Business*, a 'tween boy and girl are shown arm in arm in front of Beverly Hills Hotel. The boy's head strategically blocks the "el" in Hotel, so the word that hovers between their heads is "Hot." The girl wears a "downsized" version of a black cocktail dress; the boy wears a suit.

This kind of imagery is in marked contrast to that of Kierkegaard's day. Earlier, in 1762, Rousseau (1762/1974) had argued in *Emile* that innocence should be preserved until adolescence, by de-emphasizing gender difference until the age of puberty. His ideas about childhood innocence resonated well in the United States in the 19th century (Kaplan, 1992), providing a philosophical rationale for the visual image of the "naturally innocent child body" or "romantic child" that had already been invented in the late 18th century (Higgonet & Albinson, 1997). The image of the androgynous infant in the white dress meshed with this ideology but, by the end of the 19th century, the idea of sexual innocence was beginning to coexist somewhat uneasily with the issue of "normal" (i.e. heterosexual) child development. And, by the 1950s, US society was focused on infant gender boundaries and sexual development; a pink–blue gender dichotomy for infancy became fixed in the public consciousness. The remarkable staying power of this dichotomy is still evident in a period of industry "downsizing," in which designers ranging from Donna Karan to Tommy Hilfiger are introducing fashionable, often adultlike, apparel in smaller and smaller sizes. As if to capitalize on cultural anxieties about children growing up safely and systematically, children's apparel companies such as French Toast produce school uniforms imposed to render control and order to children's bodies. At the same time, they manufacture some of the very fashions that schools seek to displace with uniforms. Capitalism thrives on such contradictions.

Fashionable representations of children articulate the uneasy contradictions and ambiguities associated with identity boundaries of all kinds, including (a) the anxious and disorderly overlap between gender and sexuality and (b) fuzzy delineations between childhood and maturity. Questions such as the following remain open: Is it possible to maintain early rigid gender categories without emphasizing sexuality too soon? At what age does innocence end? When do anxieties about sexuality begin? Can children be modern, forward-looking subjects while also maintaining a sense of innocence?

Kierkegaard (1980) argued that innocence and anxiety coincide:

> Innocence is ignorance.... In this state there is peace and repose, but there is simultaneously something else that is not contention and strife, for there is indeed nothing against which to strive. What, then, is it? Nothing. But what effect does nothing have? It begets anxiety. *This is the profound secret of innocence, that it is at the same time anxiety.* Dreamily the spirit projects its own actuality, but this actuality is nothing, and innocence always sees this nothing outside itself. (p. 41; emphasis added)

He went on to submit that anxiety is a "qualification of the dreaming spirit.... Awake, the difference between myself and other is posited; sleeping, it is suspended; dreaming, it is an intimated nothing ... [A]nxiety is freedom's actuality as the possibility of possibility" (pp. 41–42).

Interviews with people regarding what their favorite and least favorite clothes say about their identities have revealed that it is often easier to talk about who they are *not* than it is to express who they are through language (Freitas et al., 1997). If childhood is the identity *not* of adulthood in the popular, contemporary imagination, it is apparent that this difference between self and younger other is increasingly difficult to maintain in an era of apparel industry "downsizing." Nor can the distinction be maintained in the context of media culture and the proliferation of new visual technologies (e.g., websites, cable television) with ever-increasing accessibility to individuals of all ages. As Kierkegaard (1980) prophetically noted, innocence and anxiety merge in the world of contemporary fashion culture. Children represent both of these cultural emotions, coincidentally. Fashion and everyday style signify the collective ambiguities of subjectivity that the troubled yet hopeful cultural imagination—the dreaming spirit—can express only in a visual and nonlinear way.

Higgonet (1998) argued that an irresolvable tension at the heart of Romantic childhood has become readily apparent. That is, innocence has become an object of desire. Higgonet submitted that, by the 1990s, the image of the child maintained its meanings of innocence but had also acquired others, "exchanging and mingling commercial, sexual, and political forms of power in an increasingly tight knot of private and public forces" (p. 148).

Not only may the recent "downsizing" of apparel fill the empty space between young childhood and apparent maturity; it also seems to point to the cultural anxieties about the tenuous nature of the boundaries between gender and sexuality. Perhaps culture cannot have it both ways: It cannot simultaneously insist on gender boundaries at birth and expect sexual innocence to endure until the adult years.

Compulsory gender boundaries also coincide with homophobic anxieties. These anxieties became institutionalized between the latter part of the 19th century and the 1950s, when the pink and blue symbolism became fixed. In the process of this fixing of meaning, the innocence of the Romantic child, previously depicted in an androgynous white dress, could not be maintained. But it is important to remember that the Romantic child was invented, and that the contemporary Knowing child is also an invention. It too is an invention of desire. Like innocence, anxiety cannot be separated from desire.

Part of what adults may desire are their lost childhoods as well as reproductions of themselves in the future. If this is so, this is a lot of cultural weight for children's bodies to bear. Perhaps what is needed now is more than a reinvention of visual childhood. Collective subjectivity itself requires reconstruction, or at least repair. And it may be that the visual world, especially style and fashion, is the primary vehicle for such projects in the context of technoculture.

The ambiguity of becoming is not one confined to childhood. Anxiety is, as Kierkegaard (1980) described it, an "entangled freedom" (p. 49), one that requires more than strategies of control such as school uniforms—strategies that appear to provide a "quick fix" by closing off anxieties. It is interesting to note that, just as uniforms are being placed on children's bodies, the adult workplace was becoming more casual. Symbolically, the ties middle-class men have worn are being placed around the necks of young, often minority, boys.

The entangled freedoms that constitute anxiety and, more generally, agency also include desire, creativity, and possibility. Articulating these in a way that takes multiple subjectivities—including those of children—into account begins to address what Simone de Beauvoir called "the ethics of ambiguity" (1948). She drew attention to the idea that ethics does not propose recipes, but rather suggests methods for working through the complexities that are part of everyday experience. Inasmuch as fashion and style represent entangled freedoms and ambiguous subjectivities, they offer heuristic strategies for thinking through struggles and pleasures that might otherwise not be expressed, if we take the time and energy to unravel their potential in the context of a frenetic, globalized technoculture.

REFERENCES AND BIBLIOGRAPHY

Azolay, J. F. (2001). Retail watch: Rave girls, zutopia, limited too. 'Tween Business Supplement to *Children's Business, 16*(2), 46–49.

Beauvoir, S., de (1948). *The ethics of ambiguity.* New York: Citadel Press.

Blumer, H. (1969). Fashion: From class differentiation to collective selection. *Sociological Quarterly, 10,* 275–291.

Boultwood, A., & Jerrard, R. (2000). Ambivalence, and its relation to fashion and the body. *Fashion Theory, 4*(3), 301–322.

Davis, F. (1992). *Fashion, culture, and identity.* Chicago: University of Chicago Press.

Freitas, A. J., Kaiser, S. B., Chandler, J. L., Hall, C., Kim, J. W., & Hammidi, T. (1997). Appearance management as border construction: Least favorite clothing, group distancing, and identity ... not! *Sociological Inquiry, 67*(3), 323–335.

Hethorn, J., & Kaiser, S. (1998). Youth style: Articulating cultural anxiety. *Visual Sociology, 14,* 109–125.

Higgonet, A. (1998). *Pictures of innocence: The history and crisis of ideal childhood.* New York: Thames & Hudson.

Higgonet, A., & Albinson, C. (1997). Clothing the child's body. *Fashion Theory, 1*(2), 119–144.

Kaiser, S. B., Nagasawa, R. H., & Hutton, S. S. (1991). Fashion, postmodernity, and personal appearance: A symbolic interactionist formulation. *Symbolic Interaction, 14*(2), 165–185.

Kaplan, E. A. (1992). *Motherhood and representation: The mother in popular culture and melodrama.* London: Routledge.

Kierkegaard, S. (1980). *The concept of anxiety.* (Edited and translated with introduction and notes by Reidar Thomte in collaboration with Albert A. Anderson). Princeton, NJ: Princeton University Press.

Medved, M., & Medved, D. (1998). *Saving childhood: Protecting our children from the national assault on innocence.* New York: HarperPerennial.

Rousseau, J.-J. (1974). *Emile* (B. Foxley, Trans.). London: J. M. Dent. (Original work published 1762)

Spencer, C. (1995). *Homosexuality in history.* New York: Harcourt Brace.

Wilson, E. (1985). *Adorned in dreams: Fashion and modernity.* Berkeley: University of California Press.

Winn, M. (1981). *Children without childhood.* New York: Pantheon Books.

CHAPTER EIGHTEEN

Wearing Communication: Home, Travel, Space

Patrizia Calefato

The real protagonist, the true subject and object of the late 20th-century techno-logical revolution in the field of communication, is the human body. It has be-come a fluctuating and hybrid entity, and has lost whatever it had of the unknowable and incommunicable. It is no longer a monad in search of a place, word, or gesture, but a terminal that is never switched off. The human body is en-dowed with communication, not by subjective choice, but as the result of an ob-jective condition: its status in a world where the forms of social reproduction themselves (production, exchange, consumption) have been completely ab-sorbed into communications systems. The concept of communication no longer refers merely to the transference of information from sender to recipient. The-ories of communication based on the notion of "sign" have shown that communi-cation always involves a process of interpretation; messages are reciprocal. Signs, and even signals, produced, exchanged, and consumed, imply shifts in social iden-tity, more extensive levels of connotation, myth making, and meta-communi-cational awareness.

A few years ago a TV commercial for a mobile phone company depicted a group of children vying for attention; they were seen showing off in front of their parents and teachers, making strange signs to attract their attention. The next scene depicted the same children as adults; and they were using that company's phones. The message was simple, but full of implications: Today a service offers you something that yesterday had to be sought by putting intelligence and creativ-ity to the test; namely communication—contact, recognition, giving and receiv-ing information as if we were thinking nodes in a giant network.

Communication "sticks" to the body, where it acts as both transformer and transformed. The body allows a given communication medium to dislocate and fluctuate. Today the body wears objects, not just clothes or conventional accesso-ries, but mobiles, Walkmans, global positioning systems, digital barometers, watches controlled by space stations, ski-gloves with a microchip ski-pass. An advertisement for a "communicator" shows a photo of this minute portable object, which functions as fax, email, organizer, and more; yet it fits onto a tie. The caption reads "Suits you to a T." In other words, even the most advanced "communicator" can happily coexist with the most traditional garment. We wear both, and both construct our body space as one open to total communication.

Actually, clothes have always had a communicative role. By "clothes" we mean a nonverbal language—following Barthes' definition (1998)—articulated within a *system* of traditional costume as social phenomenon (*langue* in Saussurian terminology), and a *process* of dressing as a matter of an individual's choice of attire (or *parole*). Today, fashion, or fashions, are a medium of mass communication created by the interaction between these two concepts. Yet a garment cannot be considered separately from the body, which does not act simply as its support. In the interaction between costume and attire, the "clothed body" is a structural unit of social communication (Calefato, 1999). Today the clothed body is totally communicable, like a computer interface.

In our age, the sloganized T-shirt is a metaphor for the meeting between body and language, and between system and process. The T-shirt is a "written" garment in which the body too is inscribed; it is an emblem of a garment's potential for communication, a potential that goes beyond the intention of the subject who wears it. The T-shirt is not just communicative in itself, but has increased our awareness of the fact that clothes convey meaning and that the clothed body is open to communication, in the sense of shared meanings, values, styles, and behavior.

Communication exposes the body to contact with others and extends its boundaries. Whereas the T-shirt is the meeting point between body, clothes, and language (in the form of writing), the new information communication technologies (ICTs) open up communicative possibilities for the body of a different type. They do not just introduce new media for information, but "think," program, and guide communication itself; they produce values, simulate bodily functions, and enhance organic ones.

The cyborg,[1] an invention of narrative fiction and cinema in the 1980s, illustrates the triumph of organic and inorganic accretions to the body and the growing influence of the "machine" over life, ranging from contact lenses to artificial limbs and organs, to personal computers. The suggestiveness of the cyborg lies in its representing a tangible form of science fiction: the infinite enhancement of bodily functions and the biological incorporation of a "second nature," to the point almost of mutating life—*Gattungsleben*[2]—itself. The cyborg has grotesque implications, too, the salient features of which are an emptying and opening of the body, an abasement and inversion of meaning, and a meta-stability of the body as perceptual and cognitive deception (Calefato, 1996).

---

[1]The cyborg is also the subject of theoretical writings; see Haraway (1991).

[2]In the Marxian sense; see Marx (1971, pp. 143–282).

Nevertheless, despite the validity of postulating a shift in bodily and spatial perception, today the notion of "cyborg" should perhaps be reinterpreted in a more "mature" way, like all ideas that become too fashionable and whose theoretical value is depleted by overuse. It should be abandoned, however, in the light of the experiences on which our technological "second nature" has been able to reflect.

The body thinks its own image in terms of the spatial categories that define it: high/low, right/left, shallow/deep, near/far. The cyborg dimension has called for a reexamination of the category "near/far," because the new ICTs have abolished distance and erected in its place a virtual (or semiotic) universal nearness. And the same is true for temporal relations: Email, mobile phones, video conferences, and so on are forms of communicative simultaneity, and the nearness that they occasion affects and defines the quality of the message itself. The nearness of bodies may be metaphorical and virtual, but speed is actual.

As often happens, everyday language takes on the deeper significance of the state of things that it describes. For instance, when chatting on-line, one often uses the deitic *here* to refer to the web ("How long are you going to be *here* for?"). Likewise, we now tend to give people our mobile phone number rather than our home number, keeping the latter increasingly private and, at the same time, making ourselves available anytime, anywhere, at least potentially. In ordinary conversation, when talking about communication, we refer not just to cinema and television, but to the modes of communication itself: call rates, phone models, phone–computer links, and so on. Such conversations often dwell on how intrusive mobile phones are; how their widespread, indeed sometimes excessive, use has changed our notions of public and private space; how they have made it possible for us to contact absent relatives in the midst of a family reunion; or how people may be physically present, yet distant in communicative terms, because they are talking to someone a long way away.

Thus the relation with space seems to be a new frontier for corporeality, when considered in terms of the new ICTs and the ways in which these may be literally "sewn" onto the body. The body wears communication: bits, not just atoms; signs, not just fabrics that cover it and keep it warm.

These widespread changes, however, in our perception of spatial and temporal dimensions raise certain problems with regard to the relation between subject and space. Nomad, surfer, in transit—the theoretical and philosophical metaphors used to describe the late-20th-century individual immersed in the world of the new media are drawn from the realms of space, passage, journey, movement. This imagery "frames" the condition of social subjects in our age, with all their uncertainties. Yet these same subjects are also deeply rooted in the complexity of the present, existing on both a global dimension and one of local tensions, where the body's spatial position and the maps drawn on it, which simultaneously indicate its mapping of the world, take on particular importance.

These may be metaphors of uncertainty, but they are also metaphors profoundly linked to the everyday experiences and culture of our age. The prototype and theoretical model for these subjects is undoubtedly the *flâneur* whom Walter Benjamin considered the embodiment of the metropolis; a figure of fashion and space, personifying the dialectic between open and closed spaces quintessentially expressed by the city: "Landscape, this is what the city has become for the *flaneur*. Rather, the city cleaves into two dialectical poles, opening up like a landscape before him and enclosing him like a room" (Benjamin, 1986, p. 544).

There is another image, again of citizens, in Barthes' (1984) description of Japan, which expresses the idea of transience and travel, as epitomized by objects covering the body:

> There is a profusion of what we might call "transport gear" in Japan, of every kind, shape and material: packs, bags, sacks, suitcases, bundles (the *fujo*, or peasant's scarf). Every citizen in the street carries a bundle of something, like an empty sign, energetically protected, swiftly transported, as if what were finished, concluded, in short, the illusory outline of the Japanese object, destined him/her to be an eternal wanderer. (p. 55)

Barthes is alluding here to the fact that inherently communicative objects covering the body do not just have a utilitarian function, but are invested with a mythical value that derives its force from the "stories" these objects tell. Archetypal figures, anecdotes, quotations from the various spheres of communication (cinema, literature, music), all go to make up the architecture in motion of the new *flâneur's* body.

The journey is thus inseparable from its narration: The moleskin notebooks used by Chatwin and Sepúlveda themselves figure as part of the story the travelers recount (on the inside cover of his notebook, Chatwin offered a reward to anyone who found and returned it). Today these old notebooks have become cult objects, and the tales they tell may easily be transferred to digital organizers, palm-tops, e-books, and so on.

The "venerable" Walkman and newer MP3 player mark the songlines of the metropolitan traveler. Traditional clothes take on new functions: "traveling" jackets and waistcoats full of pockets, indispensable for planetary communication, bags and trousers with special pouches for mobile phones, briefcases with compartments for floppy disks and computers, complete "portable offices" taking up just a few square centimeters.

Nomadism, "navigating" on-line by "subjects in communication," aspects that were stressed by theorists during the last decade, can no longer be separated from the idea of home: personalized commodities, personalized tranquillity, and hospitable creativity, as in the metaphor of the web "home page." A home, too, can be mobile, comprising everything that we take with us when we travel, like an office in a palm-top or a tent in an anorak. And "home" also means surrounding ourselves with a friendly environment, with familiar and personal objects: customized mobile phone holders, fabrics that regulate our body temperature, clothes that turn our movements into a source of energy, objects that keep us in plurisensorial contact with other people, whether near or far. So it is not that the body has become inorganic, as in the image of the cyborg, but rather that "things think"—objects have become humanized in order to communicate with us.

A recent advertisement for a mobile phone shows an empty Martini glass, overturned on a table, and next to it a mobile phone using a toothpick, as if it had just finished a meal. Many more images assail us every day in which objects come to life, assume the guise of humanity, and fill our homes, like "user-friendly" organic bodies.

These objects act as extensions of the present into the future, a future that is already here. They are objects without any aura of luxury or waste, because they are the result of an ever more urgent need to communicate; this communication is simultaneously contact and flux, whether a flux of information, energy, or feeling even—perhaps the feeling of a mobile phone after a good lunch. This sort of communication is consumed as soon as it is produced, and so is reproduced, reconsumed, and so on, *ad infinitum*, as in the banal example of the phone company that offers extra talk time on the basis of how much you use the phone. The ethics of recycling, too, follows the same logic: glass, ceramics, and PVC are turned into the new "technological" materials used for clothing, expressing the conscience of a society that produces waste in excess.

The age that has just passed, which Omar Calabrese refers to as the "neo-baroque," was one of stupor, based on the wonder of the new, the enraptured anticipation of the millennium, the artless admiration for things that think. This age has gradually turned into one of disenchantment. Cyberculture is now a commonplace, total communication is a physiological need, and the integration of body and machine is familiar to all. The metaphor of the home and the familiarity of "friendly" technologies is far removed from space odysseys; the personalization of communicative contexts evokes disenchantment and the end of millennium fever, of naive and prodigious expectations.

Wearing communication, putting it on our bodies, means in some sense "taming" it. And our body is still what makes us unique and determines our identity. Probably we will come to think of technology and intelligent objects maturely through a return to simplicity and discipline. There will be no more "sex appeal of the inorganic" in fashion, as in Benjamin's famous definition. If the clothes and objects that we wear "live" and "think," then we do not have to make fetishes out of them.

ACKNOWLEDGMENTS

This chapter was translated from the Italian by Lisa Adams.

REFERENCES AND BIBLIOGRAPHY

Barthes, R. (1984). *L'impero dei segni* [The empire of signs]. Turin, Italy: Einaudi.
Barthes, R. (1998). *Scritti. Società, testo, comunicazione* [Writing, society, text, communication] (G. Marrone, Ed.). Turin, Italy: Einaudi.

Benjamin, W. (1986). *Parigi capitale del XIX secolo* [Paris, the capital of the 19th century]. Turin, Italy: Einaudi.

Calefato, P. (1996). *Mass moda* [Mass fashion]. Genoa, Italy: Costa & Nolan.

Calefato, P. (1999). *Moda, corpo, mito* [Fashion, body, myth]. Rome: Castelvecchi.

Haraway, D. (1991). *Simians, cyborgs and women*. London: Routledge.

Marx, K. (1971). Manoscritti economico-filosofici del 1844 [Economic-philosophical manuscripts of 1844]. In *Opere filosofiche giovanili* [Early philosophical works] (pp. 227–249). Rome: Editori Riuniti.

CHAPTER NINETEEN

The Equipped Body: Wearable Computers and Intelligent Fabrics

Anna Poli

There is nothing new about people making use of technical prostheses and portable machines. We add these objects to our bodies in order to overcome our innate—or acquired—weaknesses, or to make the body more effective in its relationship with the world:

> The next care to be taken, in respect of the senses is a supplying of their infirmities with Instruments, and as it were the adding of artificial Organs of the natural ... and as Glasses have highly promoted our seeing, so 'tis not improbable but that there may be found many mechanical inventions to improve our other senses of hearing, smelling, tasting and touching. (Hooke, 1665, preface; also see Maldonado, 1994b)

Historically, the first wearable technical prosthesis was probably eyeglasses for long-sighted people, in about 1285 (Maldonado, 1994a). But it is certain that the first wearable machines, in the strict sense of the term, were the pocket watch (invented by John Harrison in 1762) and the wristwatch (the aviator Alberto Santos-Dumont commissioned the creation of the first one in 1907).

Wearable objects have undoubtedly changed since then. Not only are they more numerous, but they are also becoming smaller and smaller. They are now very complex, technically refined devices. Some of them even manage to integrate different electronic technologies, to the point of becoming real, powerful, wearable computers.

In many laboratories around the world where research is conducted in the field of new technologies, such as the National Aeronautics and Space Administration (NASA), the Massachusetts Institute of Technology (MIT; see Gershenfeld, 1999), and Stanford University, work has been done for the past 10 years on the design of these new wearable objects. They are computers constructed with sophisticated integrated systems, generally including a miniature screen for viewing the images generated by the computer, a miniature microphone for vocal commands, earphones to hear the computer's responses, and, in some cases, a lightweight, compact keyboard (depending on the data-entry mode employed). In the case of vocal commands, the keyboard is superfluous, freeing the hands of the user.

Thanks to continuing research, these new wearable computers are becoming increasingly simple, and designers are working on making them even smaller and less cumbersome. The term *wearable* is used to distinguish these units from portable computers and laptops. The wearable computer is so small that you can actually wear it. In recent years the continuing process of miniaturization has reached the point where computers can be made so small that they could be a part of our dress. In the future, to make them more wearable and easier to use they will be applied to fabrics or even woven into fabrics during the production of garments (Braddock & O'Mahoney, 1998).

In this field, the progress being made in the textile industry and the science of new materials has thus become very important. The new fabrics currently in the experimentation phase, as illustrated by Elda Danese in her contribution to this volume, are made of fibers whose composition is truly revolutionary. In fact, today's market already offers certain products whose fibers incorporate active substances in microcapsules. For example, the fabric of slimming pants and pantyhose contains inserts of anticellulite products that enter the legs through friction caused by movement. There are also energizing pantyhose: Thanks to a combination of an energizing active substance in the fiber and a graduated compression weave, they tonify, support, and lighten the legs. Then there is moisturizing hosiery: Legs with smooth, silky skin are guaranteed by an active substance in microcapsules integrated with the fibers. This same generation of products offers fibers that release substances to encourage sleep or combat stress, fibers that can be used for the production of relaxing pajamas or garments to help people sleep.

Nevertheless, in the specific field of "wearable technology," development is still in the initial phase. But we can safely predict that this type of "electronic accessory" will soon be widespread in the production of new hypertechnological yarns. Such yarns are designed and produced using computerized techniques. They feature fibers in which elements of digital technology have been incorporated, leading to highly innovative types of performance. These "computerized" fibers offer new possibilities for interconnection with the body as our clothing is turned into "smart" garments. It is possible that all this will lead to sweeping changes, including a transformation in our relationship with our bodies and in our way of dressing.

It is quite probable that the reference value of the garments to which we are accustomed today will change. Certainly their aesthetic value will change, along with an increase in the protective function that has always been assigned to clothing. The garment's protective capacity will go beyond the mere protection of the body from the effects of high or low temperatures, expanding toward an effective extension and enhancement of the homeostatic capacities of human beings. This will be accompanied by modification of the technical means of artificial climate control, which will be activated continuously and in a more highly personalized way. We may also see changes in posture and carriage, in the way people habitually hold their bodies in sitting and standing positions, depending on the type of clothing being worn.

Once wearable computers have become an integral part of fabrics, they will function automatically as soon as they are put on. Remaining in constant contact with the wearer, they will always be ready to supply and receive information even without specific commands. As a truly integrated system, it could be possible to control messages of different types such as email, voice messages, and text messages, while simultaneously transmitting personal data such as geographical coordinates or physiological parameters.

As we have seen, the technology involved here is characterized above all by an extreme process of miniaturization of both electronic circuits and processors. In general, a wearable computer is a calculator that is always ready to transmit and/or receive data, without interfering with other activities. The connection of the data to the computer happens by means of a network and a series of input and output devices that can change depending on the way the computer is to be utilized. Usually the signals are transmitted from one device to another without wires, by means of infrared rays, and interpreted by software programs, such as those used for voice recognition. Data and images can be visualized using different types of technology; for example, Microvision has recently developed wearable monocular retina scan displays, a new technology that makes it possible to see images in motion at the desired scale by means of rapid scanning of a ray of light that projects the images on the retina.

Some wearable computers have been designed for particular applications, such as those for the handicapped. Just consider Collins, who in 1977 created one of the first wearable computers for the blind: a digital camera worn on a tactile jacket. Such machines may bear little resemblance to other devices, although there is an increasing level of integration of sensors to monitor many physical variables.

With the introduction of sensors, it appears that the future technological development of the wearable computer will be increasingly oriented toward the more complex integrated systems known as affective computers (Picard, 1998; on artificial sensoriality, also see Mangili & Musso, 1992). These systems function without the awareness or intervention of the wearer, being activated only by emotions or sensations. They are composed of biosensors built into clothing and jewelry that are capable of monitoring affective messages through physiological signals (pulse rate, perspiration, breathing, and skin temperature), thus supplying information regarding the wearer. It may even become possible to gather and interpret personal data such as the physiological condition and/or sensory or emotional capacities of the wearer.

Such systems need to be approached with great caution. They pose dangers to individual privacy unless their output is controlled or the disclosure of one's feelings is attenuated. Why should we be interested in revealing very private information, including items that social conditioning has always taught us to conceal? Why does industry feel the need to incorporate electronic devices and new data-processing and communications technologies in clothing? Do we really need jackets lined with microchips, shoes with mood monitors, radiation-proof fibers, self-disinfecting shirts, silicone sweaters, and fabrics with therapeutic properties? And is it really true that this new generation of microtechnology will make it possible to amplify and enhance human abilities, to improve the quality of everyday life and work productivity?

If the objective of the new technologies is to design objects to solve problems and to alleviate or remove individual stress, then I am convinced that certain sections of this research will, in the future, become truly useful. I am intrigued by the possible development of new applications, especially in the area of therapeutic uses. An artificial retina already exists that will soon be able to give sight to people who are totally blind, just as cochlear implants can, in some cases, aid the deaf. And we can imagine other types of implants with implications useful for everyone, not just for people with specific handicaps. For example, thanks to bio-technology, it is possible to create certain human tissues in the laboratory, such as skin or bone.

Scientists of the Center for Future Health have designed light, compact wearable computers that should make us less dependent on doctors, by keeping us healthy and giving us more control over our own well-being. In this area we can mention health monitors, which are portable instruments for the analysis of foods to reveal the presence of allergens, and the "intelligent" toothbrush, which monitors the health of the user's teeth. Another development is the "mirror analyzer," a mirror that measures blood pressure and heartbeat by examining the iris.

The day is approaching when nanorobots will be widely utilized in microscopic electronic implants that will travel through our veins, repairing damaged tissues. But these are not the only devices that are gradually taking us toward the cyborg era, the age of the mechanically enhanced human being. In 1993, Thad Starner of MediaLab developed software to artificially increase human memory during everyday use of wearable computers: It is called the *remembrance agent*. In substance, the software automatically creates an association between a database and the images transmitted by mobile video cameras. Its advantage over a direct query–response system lies in its ability to supply information that users cannot ask for because they are unaware of its existence. For example, if you are looking for a particular address in an unfamiliar city, the software may tell you its position in relation to a department store to your right, just around the corner. Thus the software supplies information above and beyond the mere response to questions.

Continuing research in the fields of biochemical fibers and new data-processing technologies will lead to the development of worldwide solutions, perhaps through the fusion of the characteristics of multiple systems. The overlapping or interrelation of these two fields could give rise to the definition of new applications areas, from personal health care to personal monitoring and assistance. In my opinion all this could usefully enhance human functions, benefiting the personal safety, security, and well-being of every individual.

Actually we still do not know if the computer can be interfaced with human biology, and what function such a hyper-equipped body might have. But hypotheses already exist according to which a wired body could play the role of a "guardian angel" to help people live and work in increasingly complex systems (Denning & Metcalfe, 1997).

REFERENCES AND BIBLIOGRAPHY

Albini, A. (1998). Portatile, anzi indossabile [Portable, no, wearable]. In *Internet News*. Milan: Editrice Tecniche Nuove.

Braddock, S. E., & O'Mahony, M. (1998). *Techno textiles*. New York: Thames & Hudson.

Capucci, P. L. (Ed.). (1994). *Il corpo tecnologico* [The technological body]. Bologna, Italy: Baskerville.

Denning, P. J., & Metcalfe, R. M. (1997). *Beyond calculation*. New York: Copernicus Springer Verlag.

First International Symposium on Wearable Computers. (1997). IEEE ISWC'97 proceedings. CMU, MIT and Georgia Tech.

Garassini, S. (1998). *Indossa questo computer* [Wear this computer]. Milan: Arnoldo Mondadori Editore.

Gershenfeld, N. (1999). *When things start to think*. New York: Holt.

Hooke, R. (1665). *Micrografia: Or some physiological descriptions of minute bodies made by magnifying glasses*. London.

Maldonado, T. (1990). *Cultura democrazia e ambiente* [Culture, democracy and environment]. Milan: Feltrinelli.

Maldonado, T. (1994a, November 19). Gli occhiali presi sul serio [Taking eyeglasses seriously]. Contribution to the cycle of lectures "Sapere e narrare. L'uomo e le macchine" ["Knowing and narrating. Man and machines"], Florence.

Maldonado, T. (1994b). *Mondo e techne "I viaggi di Erodoto"* [World and techne in "The Travels of Heroditus"], January–April, 8–22. Edizioni scolastiche [School editions] Bruno Mondadori.

Mangili, F., & Musso, G. (1992). *La sensorialità delle macchine* [The sensorality of machines]. Milan: McGraw-Hill Libri Italia.

Mutations/mode 1960–2000. (2000). Catalogue of the exhibition at the Musée Galliera, Musée de la Mode de la Ville de Paris, Paris, April 1, 2000 [Musée Galliera: Fashion Museum of the City of Paris, April 1, 2000].

Picard, R. (1998, March 5). Affective computing: Un futuro di computer sensibili [Affective computing: A future of sensitive computers]. Proceedings of the IIMARA exhibition/conference, Montecarlo, *IMAGINA*..

Second International Symposium on Wearable Computers. (1998). IEEE ISWC'98 proceedings.

Third International Symposium on Wearable Computers. (1999). IEEE ISWC'99 proceedings.

PART IV

The Body and Technologies for Health and Well-Being

CHAPTER TWENTY

Artificial Sensory Perception: Vicarious Technologies for Synesthesia

Dina Riccò

The computers that are now invading our daily life seem to be increasingly trying to emulate human models of communication. This is not just at the level of information *representation*, from numerical codes of the first forms of human–computer communications to highly iconic figures of actual software interfaces (on the evolution of human–machine interfaces, see Anceschi, 1993), but most of all because these machines have progressively developed the acquired capacities of the *perception* of sensory data.

Before the process of computerization of many activities, the concept of a machine was as a device for the material and repeated execution of specific orders given by an operator, who was the only one endowed with all the so-called intelligent functions (see Mangili & Musso, 1992). Nowadays, as a consequence of research in the fields of science information, cybernetics, robotics, and more particularly artificial intelligence (AI),[1] the two partners—human and machine—sometimes look interchangeable.

The machine is increasingly endowed with intelligent functions, including specific perceptive skills. The machine can even boast of its own sensory organs, though they are rudimentary in comparison with human ones. It is these perceptive abilities that sometimes give us the illusion that the machine is somehow becoming human. At our disposal we have *smelling* machines, *seeing* machines,

---

[1]Marco Somalvico (1987), one of the main Italian representatives of IA, has defined artificial intelligence—in line with Minsky's and Turing's definitions—as follows: "Artificial intelligence is the computer science discipline that studies the theoretical foundations, the methods and the techniques that allow the design of digital systems (*hardware*) and program systems (*software*) that can provide the computer with capabilities that to a layperson would seem exclusively to pertain to human intelligence" (p. 11).

hearing machines; but what effects are they having on our behavior? How far does their use require a parallel mechanization of humans?

Let us take, for example, the case of automatic speech recognition, whether applied to an automatic dictation device, to an answering machine asking for information, or to any vocal interface. Its use implies a considerable change in our usual way of speaking. This is not just because it may be embarrassing to have to talk to a computer, to an inanimate object, or, conversely, because of the almost science fiction atmosphere in which we give verbal orders to a machine and then have them executed. It is especially because, in much more concrete terms, speaking to a voice recognition device requires particular attention to word articulation, pronunciation speed, volume, and so on, so as to make our language compatible with the perceptive skills of the machine.

The machine's difficulties in perceiving the human voice are less to do with hardware limitations—its highly sensitive microphones are able to receive a much wider frequency range than is available to human hearing—than with the software's abilities to enable the computer to perceive the autonomy of the words. Within a sentence, words represent morphologically distinct elements, which are however phonetically united as they are pronounced (Gozzano, 1990).[2] For the computer it is therefore difficult to understand the quality of these sounds, though only in the sense of distinguishing them and not yet of knowing their meaning.[3] Similar difficulties are present—sometimes even to a greater degree, as in the case of artificial vision systems (see Castelfranchi & Stock, 2000)—in systems related to other sensory organs.

It is obvious that what seems so spontaneous to us, such as speaking or talking, but also picking up an object or cycling, is extremely complex to a machine. To us, these actions are not the outcome of a process of calculation; in general they are not the consequence of the manipulation of symbols, but the result of a lengthy perceptive/motor training.[4] Human intelligence exploits this dual processing capability: On the one hand, the *symbolic-reconstructive* ability achieves the formation of mental constructions by processing abstract symbols (as in the case of the cognitive processes triggered by reading a written text, or any other code); on the other hand, the *perceptive-motor* ability compares the perceptive effects on us caused by every single action during the execution of some movement (see Antinucci, 1993, 1994). It is clear that the two processing capabilities closely interact, although one or the other will predominate, depending on the task. The same

---

[2]Such difficulties of distinction, though in this case related not to hearing but to sight, were present in the reading out loud of texts produced before the invention of printing, when words were written one after the other without being separated by any spaces (see Barthes & Marty, 1980).

[3]The first voice recognition system in the Italian language was created by the Centro Ricerca IBM in Rome, whose prototype was presented in 1986 with a vocabulary of about 3,000 words (see Martelli, 1987); its most evolved form, called *ViaVoice*, has now reached 64,000 words. See its website: www-5.ibm.com/it/software/

[4]Starting from this assumption, some insect robots have been created at Boston MIT; these work not in a symbolic way but in a reactive way, that is through automatic reflexes between sensory stimulus and motor reaction. See Castelfranchi and Stock (2000, p. 198 ff.).

does not happen in AI systems, which implies that in most cases it is humans that must, sometimes unconsciously, comply with the possibilities of machines.

1. SIMILARITIES AND DIFFERENCES BETWEEN NATURAL AND ARTIFICIAL SENSORY PERCEPTION

It is true—as Richard Gregory argued almost 20 years ago – that sensory organs can be considered as "transducers, essentially similar to photocells, microphones, pressure gauges, and so on … since they all turn some energy forms into signals which can be read according to a code" (1985, p. 280; my translation). However, we also know that in reality *natural* and *artificial* sensory systems operate in profoundly different ways. The reasons lie in their very nature. In central data processing there are differences at a morphological level (brain neurons are very different from computers' conductors and semiconductors); at the level of components organization (neurons are connected by a very complex network of synapses, whereas conductors' links are very simple and limited); and at the level of temporal organization (the human brain processes data from many channels simultaneously, whereas the computer does it sequentially and at higher speed). See Mangili and Musso (1992).

In the course of the history of automation, these differences have often led researchers to ask themselves whether artificial sensory perception should look like its natural model or not. The first experiments in robotics tended, in fact, to give the machines an anthropomorphic appearance. Because of the role played by cinema and science fiction production, robots were created with two eyes and a mouth, even when these were not strictly necessary to carry out the functions for which the machine was designed. This also led to the mass production of toy robots with no "intelligent" component, but whose sensory organs always had human features.

Nowadays in the field of AI, researchers prefer to speak of *emulation* of humans (Somalvico, 1992: 38), rather than of *simulation*, thus focusing attention more on the performance, though some researchers are moving in the opposite direction. Even if there is not a real return to an anthropomorphic conception of the machine, such experiments devote particular attention to the operating model of human organs.

Some artificial organs—the hand, for example—have always had their main point of reference in the natural model; others, such as the eye (see Gozzano, 1990), have now also been created, though only a few years ago it was considered impossible to build them according to the human model. An example is the *electronic eye* prototype IBIDEM (Image Based Interactive Device of Effective communication). This retinal sensor copies the functions of the retina and the iris of the human eye, with the aim of allowing deaf-mutes, in particular, to communicate face to face at a distance thanks to a new kind of videotelephony.[5]

---

[5] A description of the prototype, which is the outcome of a European research project, can be found at www.aitek.it/ibidem/ Another design for an artificial retina, though not on the natural model, is described in Mahowald and Mead (1991).

The artificial retina and the automatic ear—some of whose applications in voice recognition have already been discussed—are not the only sensory organs that have been developed for machines. The latest research into artificial perception is attempting to endow the machines with tactile skills as well. At first, all research was focused on the prehensile capabilities of the hand and on its articulatory possibilities (see the typologies of artificial hands described by Gozzano, 1990). Later on, further experiments aimed at creating an *artificial skin* that can react to the pressure exerted by the objects being manipulated. This would give the machine—in addition to its capacity to define the dimensions of an object to an accuracy of about one thousandth of a millimeter—a first degree of awareness concerning the seized object (e.g., its degree of fragility); see Mangili and Musso (1992).

Artificial olfactory sensors, like their human equivalent, are less studied. The *electronic nose* is now beginning to be researched and designs are developing. The first achievements date back to the 1990s—one of the first was the electronic nose designed at the IBM research laboratory in Zurich (see Gerber & Gimzewski, 1994)—and would appear to presage interesting applications. The most common applications concern the identification of harmful substances, both indoors and outdoors. As if to compensate for the by now acknowledged reduction in humans' discriminatory olfactory skills,[6] they are also widely used in the identification of food aromas, whereby the electronic nose can test the state of the food, or recognize the brand name of the product being analyzed. In diagnostic applications, the electronic nose permits the detection of important pathologies (e.g., tumors), without using invasive technologies.[7] It thus takes the place of the old practice of the doctor looking for any sensory evidence from the patient (see Maldonado, 1994, 1997), as in Molière's *Le Malade imaginaire* (1673).

Other uses of the electronic nose, though less widespread, are for public safety. For example, to combat terrorism at airports, they use instruments capable of detecting traces of gunpowder, which, being nonmetal, evades the metal detector. Applications of less vital importance include the *E-nose* designed by a research center at the University of New South Wales in Sydney especially to search for truffles. There are obviously some noses that are specialized for other food, especially for wines (created by Alpha M.O.S.)[8] and for olive oil (designed by the Organic Chemistry Department at Parma University).

---

[6]As G. R. Cardona (1985) confirmed: "We know that our smell threshold is generally falling, in line with some ideology of modernity which, among its various connotations, implies an aseptic attitude or even anosmia; it is evident that the bigger and bigger reference to models based on a sophisticated technology, the increasing use of neutral materials (metal, plastic) rather than traditional fragrant ones (wood, leather, bone) does not go with an olfactory involvement. We cannot bear any 'bad' smells … the right condition is therefore the *non smell*" (p. 179; emphasis added).

[7]I am referring to "Libra Nose," the electronic nose created by a group of researchers at Rome's Tor Vergata University, coordinated by Professor Arnaldo D'Amico. The principles, experiments, and applications are described at http://pendragon.eln.uniroma2.it. Other electronic noses, such as the one designed by Warwick University researchers, specialize in detecting the smell of viruses and bacteria.

[8]See *Laboratorio 2000*, November 1998; also available at www.lab2000.com/lab11-98/nas50.html

The ability of machines to pick up information from the surrounding environment has led to systems that go beyond the sensory capabilities provided by artificial organs. They are capable of detecting postures, gestures, and body movements,[9] or even of grasping their emotional content.[10]

The aspect that most directly involves us is that all these artificial senses are components that belong not only to the machine, but progressively also to humans. To some this is wonderful—Stelarc, an Australian artist with artificial arms and ears, is a case in point; to others it is also worrying. Reviving terminology that appeared to have fallen out of use, "bionic" organs are once again being spoken of, precisely to evoke this union between biological and microelectronic elements, which are more and more widely employed in building prostheses for human beings (see Zorpette & Ezzell, 2000). The artificial sensory perception of the machine is therefore paralleled by artificial sensory perception applied to humans.

2. SYNESTHETIC PERCEPTION AND MULTISENSORY SIMULATION: THE VICARIOUS POSSIBILITIES OF DIGITAL TECHNOLOGIES

One of the reasons humans are able to carry out tasks that to a machine are extremely complex is the close intersensory cooperation involved in any human activity. As Merleau-Ponty (1965) wrote, in humans "synesthetic perception is the rule" (p. 308). Our way of perceiving and processing external information works in a very interrelated and unified way (see in particular, Marks, 1978).

Although apparently simplifying the task of those designing new multisensory artifacts, because the brain already provides for the unity of the senses, this actually makes the designers' tasks more complex (see Riccò, 1999). The brain provides the unifies of not only those sensations that are already consistent with each other, but also those sensations providing conflicting information. This implies that the brain must decide which of these sensations it should attach more credibility to.

However we know, from *inter alia* Miller's experiments (1972), that in case of conflicting data the accurate sensory information is not always selected; the sensation to which the brain listens is not always the one sending back true data. For instance, when sight and touch are in conflict, giving contrasting information about the same object, sight prevails over touch, irrespective of the accuracy of the information. Therefore we paradoxically believe more in what we can see than in what we can touch. These results are confirmed by Rock and Victor's experiments (1964).

---

[9]"Elite," an optoelectronic system of movement analysis, has been designed at the Bioengineering Centre of Milan Polytechnic, directed by Antonio Pedotti (see www.cbi.polimi.it).

[10]I am specifically referring to "Eyes Web," a system for the real-time analysis of people's body movements and gestures, based on emotional agents, developed by Antonio Camurri and his team at the Dipartimento di Informatica, Sistemistica e Telematica (DIST) at Genoa University (see www.dist.unige.it, and Camurri, Richetti, and Troca, 1999). For more in general about emotional agents, see Camurri and Coglio (1998). Also see "Verso i computer affettivi" in Castelfranchi and Stock (2000: 231 ff.). On the development of experiments for the analysis of movement expression, also inspired by Rudolf Laban's "theory of effort" (Laban & Lawrence, 1947), see Camurri and Trocca (2000), as well as "Alcuni aspetti fondamentali della struttura dello sforzo" in Laban (1999, pp. 167–187).

In other circumstances, the brain cannot satisfy the demands from contradictory stimuli and consequently triggers signs of psychophysical distress, such as result from some virtual reality experiences. In some subjects, the uneasiness of the "simulator syndrome" (Pugnetti & Mendozzi, 1994, p. 52) occurs when the operations of sensory reconstruction in the simulation of a specific real context do not provide the user with a combination of sensations corresponding to the ones usually experienced in reality. More generally this leads me to argue that the attitude with which we face the problems of perception related to the new technologies will in any case have to face this peculiar interrelationship of human sensory perceptions (on the relationship between digital media and intersensory perception, see Riccò, 1996).

The scholars Grigore Burdea and Philippe Coiffet, specifically referring to systems that use virtual reality techniques, have observed the occurrence of two main phenomena of interaction:

1. *Phenomena of coordination* among the sensory faculties occur when the activation of a device requires the action of a specific organ—usually the hand or the voice—which is controlled by sensory systems that are different from it. For example, coordination is activated between voice and eye when we verbally ask the computer to move some virtual element present on a screen, simply by saying out loud "more to the right," "higher," and so on, while the results of such verbal commands, and therefore the produced sensory effects, are controlled by the eye according to what is represented on the screen (see Burdea & Coiffet, 1993).

2. *Phenomena of transposition* among the senses occur when we check the command given by a sensory channel that is different from the one of the command. Such conditions occur when the feedback from a sensitive glove is given not so much as a resistance to the gestural and proprioceptive effort, but rather, let us say, as images or sounds. The employment of such transpositions is much more common and important than one might at first imagine. In the specific case of the simulation of tactile sensations, what is fundamental—besides the actual tactile simulation, activated by contact—is the contemporary stimulation of other senses that induce tactile stimulation through synesthesia.

This is confirmed by several experiments. One of these was carried out by Burdea and his team, using a sensitive glove of their own creation, precisely in order to detect the role of visual and auditory components in tasks of tactile/proprioceptive recognition. The experiment consisted in asking subjects wearing the sensitive glove to catch, move, and change the shape of a ball in a visual simulation created using virtual reality techniques. In order to carry out this task, the subjects were aided by information in addition to the visual representation of the ball on the screen:

- Feedback from the glove itself, which responded with an intensity proportional to the force required to change the shape of the ball on the screen.

- An auditory response, whose sound frequency was proportional to the visual change in the shape of the ball.
- A visual response, presented as columns of 20 luminous LEDs each, where the number of illuminated LEDs was proportional to the ball's change in shape.

The experiment has been repeated several times both with all the sensory aids simultaneously present and with each one separately. The results have clearly shown that feedback with tactile sensations simultaneously with an auditory aid is of fundamental importance and that, in this combination, learning a virtual task occurs much faster than with either a single aid or any other combination of the ones described previously.

These possibilities of coordination and transposition in the presence of multisensory stimulations, which virtual reality systems allow and promote, can also be exploited when multisensory communication is not possible, that is, when the sensory transposition can take place only by synesthesia. I am thinking here of visual–auditory and visual–tactile translators designed for blind users, or of devices for auditory–proprioceptive and auditory–visual translation designed for deaf users. These devices exploit the potentialities of a vicarious sense in order to make up for the functions of a missing sense. In any of these devices—systems such as Braille, "the oldest example of a reading prosthesis" (Bertora et al., 1974, p. 35), which, in its most evolved form (piezoelectric Braille), translates into Braille what is on a computer screen, or computerized devices and systems that translate visual information into tactile or auditory data[11]—a specific translation code is necessary, which must be based on synesthetic principles (see Riccò, 1999).

There is no doubt that humans are synesthetic, and from some viewpoints artificial sensory perception is moving toward multisensory integration (Vito, 1994; with reference to Luo and Kay, 1989), that is, a coordinated use of the information reaching the computer from different sensors. On the whole, Simone Gozzano's thinking of some 10 years ago still seems to hold true, when he stated that the computer's senses "are still disconnected appendages" and are a long way from achieving the coordination we find in humans (Gozzano, 1990, p. 19). In spite of continuing attempts to provide machines with sensory organs that are more and more like the human ones—and sometimes with highly superior discriminatory skills, as in the case of electronic noses—generally they still look essentially *asynesthetic*. In intelligent machines we could even destroy one of the channels of perception without impairing the performance of the others, because no interac-

---

[11]One of many examples is "Optacon" (Optical to Tactile Converter), a reading system based on a sensor that detects the text of a printed page and translates it into vibration through a tactile matrix (Bertora et al., 1974). Another example is the "MOVE" (Mobility and Orientation in Virtual Environment) project—part of the European Union Technology Initiative for Disabled and Elderly People. This instrument is an acoustic virtual stick to provide blind people with precise indications for navigation and orientation in a virtual environment through the codified use of the different sounds (see Prondzinski, 1994).

tion or sensorial interference are automatically activated. However, machines, particularly digital ones, can very easily produce messages that respect the unity of our *sensorium*. It is obvious that we must take responsibility for deciding whether or not to follow them up.

ACKNOWLEDGMENTS

This chapter was translated from the Italian by Paola Iori.

REFERENCES AND BIBLIOGRAPHY

Anceschi, G. (Ed.). (1993). *Il progetto delle interfacce* [The project of interfaces]. Milan: Domus Academy.

Anceschi, G., & Riccò, D. (2000). *Research of communication design: A synesthetic approach*. Proceedings of the "Design plus Research" international conference, Milan Polytechnic, May 18–20.

Antinucci, F. (1993). Realtà virtuali [Virtual realities]. In "Editoria, applicazioni, tecnologie, virtualità," Atti del Convegno Smau ["Publishing, applications, technologies, virtuality," Proceedings of the SMAU Conference] (pp. 32–34), *Multimedia, 18*, October 2.

Antinucci, F. (1994). Il corpo della mente [The body of the mind]. In P. L. Capucci (Ed.), *Il corpo tecnologico. L'influenza delle tecnologie sul corpo e sulle sue facoltà* (pp. 17–24). Bologna, Italy: Baskerville.

Barthes, R., & Marty, E. (1980). Orale / scritto [Oral–written]. In *Enciclopedia Einaudi* (Vol. 10, pp. 60–86). Turin, Italy: Einaudi.

Bertora, F. et al. (1974). Nuovo metodo di lettura tattile [New methods of tactile reading]. *Le scienze*, no. 66, 34–41.

Burdea, G., & Coiffet, P. (1993). *La Réalité virtuelle* [Virtual reality]. Paris: Hermes.

Cagliano, S. (1986). L'uomo manipolato. Le possibilità delle nuove tecniche biomediche [Manipulated man. The possibilities of the new bio-medical techniques]. *Dossier Scienza*, no. 7.

Camurri, A., & Coglio, A. (1998). An architecture for emotional agents. *IEEE Multimedia*, October–December, 24–33.

Camurri, A., & Trocca, R. (2000). Analysis of expressivity in movement and dance. *Proceedings CIM-2000*, L'Aquila, AIMI.

Camurri, A., Ricchetti, M., & Trocca, R. (1999). EyesWeb—Toward gesture and affect recognition in dance / music interactive systems. *Proceedings IEEE Multimedia Systems*, June 7–11, http://musart.dist.unige.it

Capucci, P. L. (Ed.). (1994). *Il corpo tecnologico. L'influenza delle tecnologie sul corpo e sulle sue facoltà* [The technological body. The influence of technologies on the body and its faculties]. Bologna, Italy: Baskerville.

Cardona, G. R. (1985). *La foresta di piume. Manuale di etnoscienza* [The forest of feathers. Manual of ethnosciences]. Bari, Italy: Laterza.

Castelfranchi, Y., & Stock, O. (2000). *Macchine come noi. La scommessa dell'intelligenza artificiale* [Machines like us. The challenge of artificial intelligence]. Rome / Bari: Laterza.

De Prà, R. (Ed.). (1987). L'orecchio del computer. Il riconoscimento automatico della voce [The ear of the computer. Automatic voice recognition]. *Dossier Scienza*, no. 17.

de Kerckhove, D. (1993). *Brainframes. Come le tecnologie della comunicazione trasformano la mente umana* [Brainframes: How the technologies of the communication transform the human mind]. Bologna, Italy: Baskerville.

Gerber, C., & Gimzewski, J. (1994). Finalmente un computer che sente gli odori [At last a computer that can smell]. Interview with R. Peluso in *If, 3*, 84–91.

Gozzano, S. (1990). I cinque sensi del robot [The five senses of the robot]. *Sapere*, April, 9–19.

Gregory, R. (1985). *La mente nella scienza* [Mind in science]. Milan: Mondadori. (First published as *Mind in science. A history of explanations in psychology and physics*. Cambridge, England: Cambridge University Press, 1981.)

Laban, R. (1999). *L'arte del movimento* [The art of movement]. Macerata, Italy: Ephemeria. (First published in English in 1950.)

Laban, R., & Lawrence, F. C. (1947). *Effort*. London: Macdonald & Evans.

Luo, R. C., & Kay, M. G. (1989). Multisensor integration and fusion in intelligent system. IEEE Transactions on Systems, Man and Cybernetics. *SMC, 19*(5), 901–931.

Mahowald, M. A., & Mead, C. (1991). La retina di silicio [The silicon retina]. *Le Scienze*, no. 275, July, 34–40.

Maldonado, T. (1992). *Reale e virtuale* [Real and virtual]. Milan: Feltrinelli.

Maldonado, T. (1994). Corpo tecnologico e scienza [Science and the technological body]. In P. L. Capucci (Ed.), *Il corpo tecnologico. L'influenza delle tecnologie sul corpo e sulle sue facoltà* (pp. 77–97). Bologna, Italy: Baskerville.

Maldonado, T. (1997). *Critica della ragione informatica* [Critique of IT reason]. Milan: Feltrinelli.

Mangili, F., & Musso, G. (1992). *La sensorialità delle macchine* [The sensorialness of machines]. Milan: McGraw Hill.

Marks, L. E. (1978). *The unity of the senses*. New York: Academic Press.

Martelli, A. (1987). Parla: Il computer ti ascolta [Speak: The computer is listening to you]. *Sapere*, January, 25–30.

Merleau-Ponty, M. (1965). *Fenomenologia della percezione* [Phenomenology of perception]. Milan: Il Saggiatore. (First published in French in 1945.)

Miller, E. A. (1972). Interaction of vision and touch in conflict and nonconflict form perception task. *Journal of Experimental Psychology, 96*, 114–123.

Prondzinski, S. von (1994). Ricodificazione della realtà su canali sensoriali alternativi in soggetti disabili visivi [Recoding reality in alternative sensorial channels in the visually disabled]. In *Virtual reality in education, training and disability* (pp. 79–89). Atti del convegno [Conference proceedings], Bologna, Italy, May 13–15.

Pugnetti, L., & Mendozzi, L. (1994). Monitoraggio psicofisiologico e modelli per il recupero dei deficit cognitivi [Psycho-physical monitoring and models for the recovery of cognitive deficits]. In *Virtual reality in education, training and disability* (pp. 41–56). Atti del convegno [Conference proceedings], Bologna Italy, May 13–15.

Riccò, D. (1996). Il suono dei new media. Un approccio sinestesico ai fenomeni d'interazione sensoriale nei software multimediali. PhD thesis in Industrial Design, Milan Polytechnic.

Riccò, D. (1999). *Sinestesie per il design. Le interazioni sensoriali nell'epoca dei multimedia* [Synaesthesia for design. Sensorial interactions in the age of multimedia]. Milan: Etas Libri.

Rock, I., & Victor, J. (1964). Vision and touch: An experimentally created conflict between the two senses. *Science, 143*, 594–596.

Somalvico, M. (1987). La realtà scientifica dell'intelligenza artificiale [Scientific reality of artificial intelligence]. *Media Duemila*, no. 42, 8–23.

Somalvico, M. (1992). Il vero, il buono, il bello del bipolo uomo-macchina [The real, the good and the beautiful in the double pole man–machine]. *Media Duemila*, no. 93, 33–49.

Various authors (2000). Futuro bionico [Bionic future]. *Le scienze dossier*, no. 4.

Vito, R. (1994). Percezione [Perception]. In O. Stock (Ed.), *Intelligenza artificiale* [Artificial intelligence] (pp. 93–114). Milan: Franco Angeli.

Zorpette, G., & Ezzell, C. (2000). Il futuro bionico [The bionic future]. *Le scienze dossier*, no. 4, 4–5.

CHAPTER TWENTY-ONE

Health Care Technologies:
The Contribution of Industrial Design

Medardo Chiapponi

My discussion in this chapter focuses on technologies aimed at the maintenance and restoration of the body's well-being. More precisely, I look at certain types of technology, such as health care equipment, medical scientific instruments, and equipment for analysis, diagnosis, therapy, surgery, rehabilitation, and support.

In order for a discipline such as industrial design to be applied to such themes, we need to consider how industrial design can contribute to increasing the well-being of the human body. That is, we need to understand how the discipline can use its own methodology and instrumentation to improve the performance—in the widest sense—of health care technologies.

Potentially, industrial design can raise the qualitative level of prevention, diagnosis, care, and rehabilitation services offered to the public, as well as improve the working conditions of medical, paramedical, and technical personnel. However, its actual ability to do this is directly proportional to the proper use of some of the discipline's particular characteristics.

In this regard, the user must always be a central focus. Consideration must be given to human and sociocultural factors in the design of products. In other words, particular attention must be paid to what the "usage value" might be of any features. Such accentuation of the aspects connected with usage value sets industrial design apart, as opposed, for example, to design activity based on engineering, which generally focuses on "technological value" or "economic value." The need for user-oriented design is painfully evident in the field of health care, where otherwise advanced technological instruments and equipment overlook psychophysical and psychological needs, which are essential components of the well-being of the users of the equipment.

Contradictions of this type seem to be appearing more often. One reason for this is the difficulty in determining what might be included in the concept of "the user." In planning discussions the user often becomes transformed nonchalantly from a physical person into a statistical entity. By contrast, industrial design focusing on the user must represent the interests of a future user in a collective decision-making process, act in the name of users, and respond to an implicit assignment to confront and resolve problems related to the product system.

At this point, however, at least in industrial design discourse, the focus of the debate has shifted; some have even challenged the very idea of this implicit mandate. The tendency is to emphasize the role of individuals as their own "personal designer," restricting the range of action of the specialists. Some theorists would urge individuals to transform themselves from mere users of the products offered to them into designers of their own "product milieu" (Chiapponi, 1999). Without becoming naive "do-it-yourselfers," individuals may be motivated by various reasons and circumstances to address these problems personally: greater possibilities than in the past to have an effect on product policies; greater attention to users on the part of manufacturers; a wider spread of specialized knowledge; relatively easy access to sophisticated technologies.

This situation seems to blur the distinction between "functional knowledge" and "structural knowledge," which used to be decisive in preventing the passage from the condition of user to that of designer. In effect, from this point of view there has always been a significant difference between one user and another, and one product family and another. This gap depends, above all, on the different levels of specialization of both users and products. Scientists, for example, have never been passive users of their working tools, but have always taken an active part in their design and realization. Some of the greatest successes of science are linked to the appearance of new, more powerful instruments, with scientists making a decisive contribution. This process extends to health care, where physicians and other specialists may be the forerunners, on an experimental level, of the wide-scale production of instruments and equipment.

The fact that users are capable of establishing an increasingly effective dialogue with design professionals does not, however, exonerate the professionals from continued attention to the user. On the contrary, the professional skill of industrial designers can be applied to increase the autonomy of users by, for example, transferring technologies initially introduced in highly specialized sectors to systems of products whose utilization can be extended to non-professional users.

A theme such as that of health care is ideal, in many ways, for industrial design. As I mentioned earlier, the industrial designer operates as the interpreter of users, the connecting element between the user and the manufacturers that ensure the production and distribution of products. In this perspective, the designer's main task is to construct a system of plausible relations between the system of products to be designed and a corresponding system of needs. The field of health care equipment is an ideal field of application precisely because the dynamics behind the relationships among needs, requirements, products, and services are more favorable. This situation of advantage is essentially due to a clearer identification of users (patients and health care professionals) and a direct relationship between these professionals and manufacturers. Therefore, mediation is easier because of the more precise definition of the characteristics and needs of those who will use the products, and the direct, competent participation of the users themselves in the phases of design specifications, testing, and experimentation. Another element of greater clarity stems from the fact that the prevalently functional character of health care products, and the clear identification of their users, makes it easy to establish the necessary priorities among all the parameters and factors to be considered in the design process.

The contribution that industrial design can offer toward the health and well-being of the body is moving in the direction of greater concreteness. In the industrialized nations, health care is the subject of heated debated on, among other things, the relationship between public and private organizations, the type of services offered and their distribution in the territory, the public budget allotments for health care, and the financial burden assigned to individual users. These discussions undoubtedly address themes of great importance, but also risk producing results that fail to meet expectations, because they are often conducted on an abstract plane. In design practice, because we have to come to terms with technologies, objects, and machines, that is, with the solution of concrete problems, the results become immediately perceptible, to the extent that the projects developed are capable of effectively increasing the qualitative level of the services of prevention, diagnosis, and care provided for citizens.

The pragmatic character of industrial design and the concreteness of its aims can be seen in its very definition, formulated in 1961 by Tomás Maldonado and adopted that same year by the International Council of Societies of Industrial Design (ICSID), the leading professional organization in this field on an international level. According to this definition, industrial design has the task of designing the form of industrial products, and this means both "factors related to use, to individual or collective consumption of the product (functional, symbolic or cultural factors)" as well as those related to its production (Maldonado, 1991, p. 12).

The aim of industrial design is to "give form" to products, or to create a synthesis of a wide variety of aspects (technological, functional, performance related, formal aesthetic, etc.) while deciding, case by case, which of these aspects will have the greatest priority. In practice, "giving form" to products means, in this perspective, contributing to making the solution to the different problems that arise in the sphere of material culture concrete, materializing the solution in its details—in our specific case, to concretizing solutions for the problems of health and well-being of citizens.

This definition leads to other important consequences. One is the natural propensity for interdisciplinary work, seen both as a "synoptic" way of working, taking multiple factors (formal, performance related, technological, economic, etc.) into account in each project, and as an inclination to cooperate with many other professionals. A further consequence of the attention paid to the different factors that determine the form of products is the need to be informed regarding transformations in neighboring sectors. In some ways we can even assert that today the most notable variations in design are caused by changes taking place in the contexts in which it operates, especially the technological, sociocultural, economic, and environmental contexts. Consider, for example, the effects of sociocultural transformations that modify individual and collective behavior, leading to an increasingly important role for citizens in everything that has to do with their health and well-being, including the choice and use of technologies and products.

Even more important for our purposes are technological events of great impact, such as progress in the sectors of materials or of laser and ultrasound technologies, or the spread of microelectronics and, more generally, of information and telecommunications technologies. The information technologies have opened new paths for both machine tools and production processes, and for the products themselves. The most evident of these trends are extreme miniaturization and reduced energy consumption. Another effect of microelectronics is the elimination of the monofunctional character of single components (in contrast with mechanical and electromechanical products). In simplified terms, microelectronic products contain components—microchips—that perform highly differentiated tasks, enriching the performance characteristics of each product.

The combination of these different technological innovations permits aggregation in a single product of previously separate performance functions, thus providing a stimulating opportunity to profoundly modify the established product typologies of our material culture, giving rise to completely new ones. The general public is familiar with these trends in popular consumer goods (e.g., home entertainment equipment), but far more significant are the changes in products with a high level of functional value, particularly in the field of health care equipment. To cite just a few examples, we can mention the progress in the field of image diagnostics, the introduction of nanotechnologies in the sector of internal prostheses, and the birth of video-laparoscopic surgery thanks to new products that combine diagnostic and operative functions.

All the aspects just discussed can be organized into four main working areas that express thematic emphases and the capacity to respond to specific problems: the design and planning of products, the design of services, the design of environments, and products and communication.

1. THE DESIGN AND PLANNING OF PRODUCTS

A central objective of design in the field of health care is to emphasize the role of products in the interface between health care professionals and patients. This means increasing the usage value of the products by enhancing their operative and performance effectiveness, but also by making them easier to use and transforming them into instruments of diagnosis, care, and rehabilitation that pay closer attention to the delicate physical, psychological, and existential conditions of the patient.

Industrial design can also play an important role in favoring more direct contact between producers and users, getting users involved in the definition of the characteristics of products and in experimentation. In this sense, a further task is the reinterpretation of the needs expressed by health care professionals, in the form of functioning prototypes or independently constructed "one-offs," for example.

It is also necessary to participate in the planning of health care products in the different locations where it takes place, from manufacturing companies to the organizational structures of the health services that procure the products required for their work. In the product planning conducted by manufacturing companies, industrial designers rely on specific project-oriented expertise; but when they are operating within the organizational structures of health care services, analytical capabilities are equally important. Besides offering indications for the design of new products and actually designing them, the industrial designer must compare and select existing products, expressing circumstantiated judgments on their compliance with existing standards, their performance, their ease of assembly, use, and maintenance, the durability of the complete product and its components, the quality–price ratio, and so on. Therefore the designer's contribution also involves the expression, in analytical, systematic, and technically reasoned terms, of the evaluations of products articulated by potential purchaser-users in a more synthetic manner, without all the necessary information and knowledge.

2. THE DESIGN OF SERVICES

The contribution of industrial design in the design of health care services (services of prevention and diagnosis, and the management of the life cycle of medicines in the hospital—supply, storage, distribution, disposal, etc.) is, in many ways, similar to the contribution made in product planning. Again, close cooperation with the health care professionals is essential, especially with those who plan the service from an organizational, logistical, and economic viewpoint and with those responsible for the management of the service itself. At the same time, the relationship with important external figures such as manufacturers is also fundamental. The designer's particular task is first of all to establish a system of products (selecting products already on the market, designing and making new ones, eliminating obsolete or useless products) that can contribute in the most effective possible way to the achievement of the general objectives.

It is evident that the efficiency and indeed existence of health care services require suitable products, in addition to an adequate organizational and managerial structure, qualified personnel, and other properties. The furnishing of hospital services would be impossible without the coordinated assemblage of electromedical devices, diagnostic equipment, instruments for traditional or innovative surgery, equipment for emergency room services, scientific instruments for laboratory analysis, structures and computers for the management of medicines, furniture for in-patient wards, and thousands of other products. A key condition of the efficiency of the service is that these different systems of products be planned, eliminating superfluous and obsolete items, keeping up with technological innovation and the evolution of clinical knowledge, ensuring functional quality through maintenance and repair plans, establishing the need for products to be created if they are essential and not yet on the market. This is the type of operation in which industrial designers can work with ease, thanks to their professional training.

3. THE DESIGN OF ENVIRONMENTS

The micro-environmental quality of particular types of hospital spaces (operating rooms, intensive-care units, spaces for the care of newborn babies and for nuclear medicine, outpatient clinics, inpatient wards, spaces for relatives and visitors, etc.) and nonhospital facilities is key to the achievement of a high level of health care service. Expertise in the field of design and analysis of products and systems of products enables industrial designers to collaborate effectively with other specialists to achieve these quality objectives.

More specifically, industrial design can make a particular contribution to a phase of the design process that does not appear to be adequately addressed today, in spite of its importance. This is the architectural design and construction (or restructuring) of buildings, right up to the stage of concrete functional implementation and activation. During this phase, the equipment and infrastructures required for effective performance of the services are selected and installed. For the design to be effective, it is decisive in this phase to seek the integration of the equipment and the infrastructures in the micro-environment.

The performance potential of the equipment can be fully exploited through the accurate design of its systems configuration. In other words, the designer works on the physical, functional, and perceptual connections among the elements of the system of products utilized in a given micro-environment, and the connections between the products themselves and the architectural container. The objective is to increase the quality of the service provided to patients, while improving the working conditions of the health care professionals. This approach also facilitates a proper focus on the needs for flexibility and transformability typical of health care facilities and hospitals in particular.

4. PRODUCTS AND COMMUNICATION

The importance of communicative aspects in the design of health care products essentially results from the widespread presence of digital technologies in this field, and the great structural and functional complexity of much of the equipment used for analysis, diagnosis, care, and rehabilitation. Profound changes have taken place and continue to take place in the modes of designing, producing, and utilizing such equipment. A progressive increase in indicators and information devices in products (counters, readouts, visual and acoustic signals, etc.) with respect to control devices (switches, buttons, levers, knobs, pedals), which enabled an enormous growth of performance possibilities, coincided with a general reduction in the size and weight of products. Of all the forms of communication that accompany the entire life cycle of products, two are particularly relevant in this context: communications interfaces and instruction manuals for use, maintenance, repair, assembly, and disassembly.

Obviously the design of the form of a product has always involved its communicative interfaces. But the spread of digital technologies has enormously increased the importance of this design theme. In the field of health care, for example, it is becoming essential to design not only the whole system of equipment utilized to effect refined diagnostic studies using images, but also the interface through which the results of these studies can be effectively made available. Otherwise, the possibilities for fully exploiting the capabilities offered by such highly sophisticated products would be drastically and unjustifiably limited.

The design of instruction manuals is equally important for the optimal functioning of a product or a system of health care products. This facet of the product–communication relationship has affinities, and therefore possible synergies, with the design of communicative interfaces. For products involving digital technologies, in fact, the instructions are often "on-line" and their design, in practice, is connected with the design of parts of the communicative interfaces of the products. However, instruction is a crucial part of industrial design in the health care sector, and therefore cannot be limited to the design of on-line help systems; it must also include manuals that exist in a form separate from the product itself (e.g., printed texts, audio cassettes, CD-ROMs).

My subdivision of the possible interventions of industrial design in health care into four areas is meant to be only illustrative. In practice, designers are faced with problems that require a combination of these different types of intervention. Nevertheless, because I am proposing the introduction of new disciplinary contributions to help solve important problems in health care, it is necessary to be clear about which cases might benefit significantly from such contributions, just what the contributions might consist of, and how they can be integrated with those offered by other professionals. This is precisely the aim of the examples utilized here—to indicate the possible contribution of industrial design in the field of health care technologies.

REFERENCES AND BIBLIOGRAPHY

Chiapponi, M. (1999). *Cultura sociale del prodotto. Nuove frontiere per il disegno industriale* [Social culture of the product. New frontiers for industrial design]. Milan: Feltrinelli.

Maldonado, T. (1991). *Disegno industriale: Un riesame* [Industrial design. A re-examination]. Milan: Feltrinelli.

CHAPTER TWENTY-TWO

The Third Skin:
Wearing the Car, Ignoring Safety

Jorge Frascara

The car maybe represents the most removed outer skin we can wear. As such, it allows us to acquire personalities that are quite different from our real ones. With help from advertising narratives, the car—beyond transportation device—has become a part of our body, drug, and social communication. Perceived as protection and as expression, the car sometimes invites us to act at the edge of our personal safety, to the point of having become the most dangerous illness for people under 35 years of age. (World Health Organization, 1973; cited in Naatanen & Summala, 1976, p. 13)

A third skin. The car as apparel. Wearing a 1,500 kg magic costume that transports us at 100 km per hour creates interesting problems of adaptation. Allow me to explore them:

- The phone rings. The hospital calls. They tell you that your brother has been hit by a car and is unconscious. They want you right away.
- Another day. You are standing on the sidewalk, waiting for the green light in order to cross. A car speeds by, trying to beat a red light. The driver loses control and the car hits a lamppost. The driver is thrown out of the car and lies on the pavement, moaning. You go to help. It's a mess. You want to throw up. You feel faint.
- It's Saturday. You are returning from a disco with your mates. The driver is happy and is going fast. Everybody is having a great time. You don't say anything, you just want to get home. It's up to you to write the end of this story.

In Italy, every 2 minutes an injured person is taken to hospital after a traffic accident (*Giornale dell'Ingeniere*, 2000). Every 2 minutes a stupid or distracted driver does something silly that injures someone. This is not to be seen as statistics; the numbers don't count. Pain does. Lots of pain, and joy lost, sometimes for the rest of the life of the victim, if there is life left. In fact, in North America, every hour, 20 survivors of traffic collisions leave hospital marked forever: missing a foot or a hand, disfigured, or in a wheelchair (this is extrapolated from statistics from Alberta, Canada, which represents 1% of the population of Canada and the United States combined; see Alberta Transportation and Utilities, 1990; see also National Safety Council, 1992).

During the Vietnam War, cars killed 10 times as many Americans at home as fell in Vietnam. No one protested about the tragedy on the streets in the United States. It is accepted as if it were natural death. We must, however, wake up. It is an intolerable situation. Once one knows about it, it is no longer bearable.

In general, 97 percent of the population drives well enough to stay out of trouble. It's necessary to reach the other 3 percent, but in terms of the people, not the numbers. Numbers don't suffer. Numbers don't bleed. My brother does. My friend, my spouse, my child.

Let us look at how this drama of the third skin gets constructed—the car as an extension of the body, which expresses us, protects us, and invites us to act in the public space.

The car is not just a means of transportation; it is also a means of social communication. We wear the car like a costume: For some of us it's a warrior's armor; for others, their team shirt; for others, a sexy expression of their lifestyle.

From the day Henry Ford understood that the one-model, one-color approach to car making would not succeed in the market, advertising began to feed people's fantasies and to change the car from a transportation device to a communication device. The car has become a stronghold to which to anchor fantasies, and advertising constantly creates them.

There is the fantasy of the female in love who waits for her lover at the seaside, gazing at the horizon hoping to discover the familiar sailboat, standing beside her luxury sedan car. So the car is loaded with the sea, the horizon, infinite love, luxury, sailing boats, leisure time, freedom ...

There is a man with his arms around two women, standing behind a Pontiac, while kite fliers pass by, and the text reads: "Pontiac. We build excitement" ...

There is the man watching the sun go down behind the hills, standing in an open Jeep. "Only in a Jeep," the text reads. We are confronted with solitude, freedom, wide horizons, nature, fundamental human needs for prospect and refuge (well known in the literature on perception psychology). The notions of prospect and refuge come from a theory initially proposed by Appleton (1975), which suggests that people enjoy situations in which there is a panorama in front of them, while their backs are covered. He argued this is connected to the human experience of thousands of years living outdoors and being concerned about safety.

Then for those not impressed by nature there is the technological discourse. This rhetoric is supported by seemingly technical drawings, photographs of engines, and expressions such as "Inside, 39 percent more power" (BMW), as if 39 were substantially different from 38 or 37, and actually meant anything. Here we encounter the myth of precision, a common argument when discussing German engineering.

There is also the juxtaposition of racing cars with city cars (Ford); on a magazine cover (*Car and Driver*): "At 179 mph all you hear is the pounding of your heart." It is no surprise that the car in question (a Chevrolet Corvette) is one of the cars with the most collisions per unit sold in North America (Edmonston, 1990).

The fantasy world created around the car knows no bounds. The fantasy of policemen's courage links a 4x4 to frontier police (Suzuki). There is also, for secondhand cars, the criminal's fantasy: "Druglord 4x4s $100" reads an ad in *Village Voice*.

Then there is mud. A 4x4 covered in mud is given a dual personality—half clean, half muddy—with the slogan "Half man, half beast" (Nissan Pathfinder). For the more timid mud-loving cyclist, it's possible to buy a T-shirt with a mud-spray pattern printed in different colors, so that one can appear dirty while being perfectly clean.

The 4x4s abound in metaphors. In specialist magazines one sees exercise machines advertised with bodybuilders as models, with the aim of transferring the illusion of those bodies to car buyers. (You might not have the body, but you can buy the car!) Some years ago, two brothers drowned inside their 4x4 in a Canadian river, maybe having seen too many advertisements in which 4x4s had no problem in crossing rivers.

There is a generalized resistance to putting the blame on the car. A Canadian newspaper published the picture of a serious collision on its front page under the caption "Killer Roads!" (*The Edmonton Sun*), as if the roads were to blame. The roads are infrastructure provided by the government, whereas cars provide funds to newspaper businesses through advertising. No one wants to attack the cars or the drivers. It's much easier to blame the government. The other easy target for blame is the weather: "Bad weather, chain of collisions in the city" (*Corriere della sera*, October 15, 2000), as if drivers were unaware of the rain and ignorant of the fact that it's different from driving in perfect conditions.

In car advertising, safety appears to be associated with devices inside cars: seatbelts, airbags, collapsible fronts, and so on. Little mention is made of the fact that none of this helps pedestrians, who make up a high percentage of traffic casualties. Moreover, these safety devices do not contribute to reducing the main causes of collisions, which in Italy are distraction (17.8 percent), excessive speed (12.2 percent), following too close (11.4 percent), ignoring the right of way (11.0 percent), and going the wrong way (2.0 percent) (*La Repubblica*, October 20, 2000, p. 37).

Newspapers often publish pictures of collisions, particularly spectacular ones with lots of twisted metal. They may be trying to promote caution in driving, but this doesn't impress anybody; it only entertains readers and makes a spectacle of accidents. According to research in Australia (Advertising Federation of Australia, 1990), the crucial argument to be used in raising people's awareness in relation to driving is not twisted metal, or even the possibility of one's own death—in which the young don't believe. There is more responsiveness to the possibility of being disfigured or paralyzed for life; even more believable is the danger of being responsible for the killing or the maiming of a loved one: lover, relation or friend ...

Everything communicates, but communication emerges from cultural constructions, not from the objects themselves. When the hockey helmet was introduced in Canada, some players did not want to use it, regarding it as something for weaklings. Today, after universal acceptance, the helmet has become a symbol of manliness.

The paths of communication sometimes become very convoluted. Red cars suffer more collisions and receive more speeding tickets per unit sold than cars of any other color. The police know this, and pay more attention to red cars, thus reinforcing the statistics. This does not affect the Ferrari Testarossa, however. It communicates power, speed, and luxury, even when parked. Its owners apparently do not need to go fast in order to express what they want to express.

On another dimension, there is also the myth that large cars are safer than small cars. The size of the car, however, has very little to do with safety. The Chevrolet Camaro, which is similar in bulk to a Volvo Station Wagon, is eight times as dangerous (Edmonston, 1990). It is not so much that the Volvo is safer than the Camaro; it is that the buyer of a Volvo is very different from the buyer of a Camaro.

In 1987 the Canadian government recognized that traffic collisions were a major health and safety problem in the country. In the previous decade, in a population of 30 million, 51,300 people had been killed and 2,342,300 had been injured in traffic collisions, many of whom were marked for life. For people under 45 years of age, traffic collisions were the major cause of death and an important cause of injury (Road Safety and Motor Vehicle Regulation Directorate, 1987). In addition, few resources were dedicated to research into traffic safety, in comparison with funds supporting cancer and heart research. In the United States, in 1987, injury research of all kinds received U.S.$160 million, whereas the National Cancer Institute received U.S.$1.4 billion and the National Heart, Lung and Blood Institute got U.S.$930 million for heart research (Rice et al., 1989; see also Frascara, 1997).

Between 1995 and 2001 the number of cars in Italy rose by 5 percent. Traffic accidents, in contrast, increased by 47 percent. Today it is estimated that the cost of collisions in Italy is 22.2 million euro per year (*Giornale dell'Ingeniere,* 2000).

The solution proposed by engineers is to build safer roads; the police propose more personnel. Such complex systems require caution however (for a good publication on complexity, see Bocchi & Ceruti, 1992). If a narrow road is widened to make it safer, drivers will go faster, canceling out the improvement. It is necessary to build safe roads, but it is also necessary to train drivers, and, often, also to increase the number of police officers.

The car envelops the whole body, symbolizing it and protecting it like a third skin. With the help of advertising, it allows us to live out our fantasies and communicate with others. In Alberta, in Canada, after 13 years of research and 4 of public action, in 2000 there was a 19 percent reduction in traffic-related deaths. There are thus 78 people walking around Alberta who would have been buried if it had not been for our work. Although I regard this communication design job as inferior to others oriented to more sophisticated levels of experience in life, at least we have managed to keep more people alive and able to enjoy their bodies.

Just as children hurt themselves when playing because they are not aware of the fragility of the body, adults, playing with cars and pushed by advertising myths, risk their real skin, ignoring the fact that, at 100 km per hour, the hardness of a car becomes soft if it hits a wall, and all the myths can be destroyed in a second.

We must oppose the myths broadcast by advertising, and promoted by all the devices we can wear, with an awareness of our fragility. We must also wake up to the preciousness of our treasures of sight, touch, and physical activity, of the mind, and of the joy of friendship and love. Our natural body is far superior to anything we can wear.

ACKNOWLEDGMENTS

My thanks to Leopoldina Fortunati and Raimonda Riccini for their invitation to link my research on traffic safety to the scope of this project.

REFERENCES AND BIBLIOGRAPHY

Advertising Federation of Australia. (1990). *Effective advertising. Casebook of the AFA advertising effectiveness awards.* Sydney, Australia: Author.

Alberta Transportation and Utilities. (1990). *Alberta traffic collision statistics.* Edmonton, Alberta, Canada.

Appleton, J. (1975). *The experience of landscape.* London: Wiley.

Bocchi, G., & Ceruti, M. (1992). *La sfida della complessità* [The challenge of the complicit]. Milan: Feltrinelli.

Edmonston, P. (1990). *Lemon-aid used car guide.* Toronto: Stoddart.

Frascara, J. (1997). *User-centred graphic design: Mass communication and social change.* London: Taylor & Francis.

Giornale dell'Ingeniere. (2000). "Un vero bollettino di guerra" [A real war bulletin]; and "L'Italia e un obiettivo: migliorare la mobilità" [Italy and a goal: Improve mobility]. no. 19, November 15.

Naatanen, R., & Summala, H. (1976). *Road-user behaviour and traffic accidents.* Amsterdam: North-Holland; New York: American Elsevier.

National Safety Council. (1992). *Accident facts.* Itasca, IL: Author.

Rice, D. P., Mackenzie, E. J. et al. (1989). *Cost of injury in the United States. A report to Congress.* Institute of Health & Aging, University of California, San Francisco, and Injury and Prevention Center, Johns Hopkins University, Baltimore.

Road Safety and Motor Vehicle Regulation Directorate. (1987). *Smashed.* Ottawa: Transport Canada.

CHAPTER TWENTY-THREE

Conditions of Microgravity and the Body's "Second Skin"

Annalisa Dominoni

Continued manned presence on space stations or Earth-orbiting pressurized modules requires research on habitability issues for human beings in confined microgravity environments. This research is necessary to support sound industrial design strategies for comfort and functionality and, therefore, increased efficiency and better working conditions in outer-space habitation. The task of industrial design is therefore to design instruments, materials, and tools, including clothing, that facilitate human movement and the carrying out of the different activities required of space station inhabitants. In terms of living environments, the aesthetic requirements of the user/astronaut must be met while also ensuring conditions of habitability that are truly stimulating. These should not only reflect a sense of attractiveness but, even more important, be able to re-create artificially much of the Earth's environmental stimuli (light, air, odors), which are not ordinarily present in confined microgravity environments. Such efforts are of course aimed at enhancing the psychological, social, and physiological well-being of the crew.

The industrial design research project for an intravehicular activities (IVA) integrated clothing support system (VEST), discussed in this report, entails the development of an integrated clothing system for space crew. The project, proposed by me (Head of Research Programs at SpaceLab, the space design laboratory at the Polytechnic of Milan) and sponsored by the Italian Space Agency (ASI), is closely involved with the issues for the International Space Station (ISS), which is currently being built in the laboratories of 16 countries and assembled in space. Owing to its conditions of microgravity, the space station will be ideal for conducting biological and advanced materials research.

Eventually, the crew will no longer consist exclusively of solidly trained and experienced professional astronauts, but also will include research scientists and even tourists. These latter individuals are not prepared physically or psychologically for the extraordinary rigor of long-duration space flights or for living in constant harsh conditions.

Industrial design needs to consider the factors that are essential to a stimulating, attractive, and efficient environment for human beings in outer space. The demand for this research arose from the conclusions of the National Aeronautics and Space Administration (NASA) studies on the psychological and physiological needs of the crew associated with an increase in the length of space missions from a few weeks' duration to 6 or 12 months, if not longer, on the ISS. Prior to the studies, the human factors of the crew were neglected and viewed as secondary in comparison with other priorities; now, however, the conclusions have led to a form of design aimed at increasing the comfort and mental well-being of the individual (Dominoni, 2000b).

Because the primary goal of securing people and material in orbit has been achieved, the focus is now shifting to living conditions. The engineers, physicists, chemists, biologists, and doctors who have predominated in space operations will in the future be joined by an increasing populace including psychologists, sociologists, architects, and designers. Living in space entails having to deal with, among other exigencies, microgravity, a confined environment, and cosmic radiation.

First, microgravity (or partial gravity) is at the moment considered the greatest obstacle and the most difficult one with which to deal because it impairs the physiological, nervous, cardiovascular, and muscular–skeletal systems. Once in orbit, the crew find that many things are changed. Their posture, their movement within the station modules, and their sense of balance are altered, as are their sense of direction and perception of space, volume, and colors. Their body size and shape change: The spine lengthens, leg volume is reduced, and fluids shift from the lower to the upper body, causing the "puffy face" phenomenon (Antonutto, Capelli, & di Prampero, 1991). The sum of these processes is a total change in body shape. Second, the confined and isolated environment leads to emotional stress and an increased need for privacy (Bluth, 1985). The admixture of multiple ethnicities, cultural norms, and gender issues exacerbates aggressiveness among the crew. The problems are made worse by the omnipresent sense of risk and remote supervision from ground control (Clearwater, 1990a). Third, the crew are exposed to cosmic radiation that is 1,000 times higher than that on Earth.

1. OBJECTIVES OF THE PROJECT

The most serious impediment to living and working in space is reduced gravity's damaging effects on the body. A truly innovative aspect of this research project is the idea of turning this limitation into an advantage, that is, benefiting strategically from the features peculiar to the microgravity environment.

The research project sketched in this chapter is part of a larger project developed by me inside SpaceLab. The objective of the project is to investigate the dynamics of human movement in conditions of microgravity and to probe the relationship between body and objects, regardless of the context. The aim is to make equipment more flexible and suitable for both the modules of the ISS and the interior of launch vehicles, as well as for shuttles for future manned missions to Mars.

Anticipating the use of intelligent fabrics and the insertion of prostheses to aid mobility and facilitate the anchorage of operators to the internal structure of the ISS has led to the development and design of an integrated clothing system for astronauts. Industrial science researchers and designers are working in cooperation with Benetton in an attempt to create fabrics suited to the specific features of life in space, and they are producing prototypes of clothing with designer Alenia Spazio for testing in space and project validation.

The first step in meeting our objectives is to identify the main requisites of fabrics and garments to be used on board the ISS so as to develop a new integrated clothing system capable of enhancing body mobility and making the performance of crew activities more efficient and comfortable.

At present there is no specific clothing for use in space that can compensate for the many physical discomforts of life in a confined microgravity environment. These discomforts include wide variations in body temperature, lack of privacy, enormous time requirements for fitness regimes to retain bone and muscle mass, and increased perspiration.

Moreover, many astronauts would prefer to wear clothes that do not smack of military uniforms, but instead enhance their personality and identity. A positive self-perception includes feeling at ease with one's body and one's clothing. Consequently, our study of fabrics, shapes, styles, accessories, and colors is aimed at the diversification of garments to offer astronauts a wider choice. This line of investigation should yield improvements in the astronauts' feelings of well-being, quality of life, and onboard social relations. Clothing truly comes to be viewed as a "second skin," able to meet the functional needs of body and spirit alike.

The study may well lead to the creation of a variety of clothing and underwear suited to the multitude of activities on board. However, the design of such clothing will also have to take into account the problems of the onboard storage of material; packaging will need to be as compact as possible.

Another problem that needs to be addressed is the maintenance and, possibly, the washing of the garments themselves. Clothing is currently packed and brought back to Earth unwashed. Research coordinated by SpaceLab is being carried out to develop a washing and drying system for clothing on board the ISS. The research project recommends that Electrolux Zanussi be engaged in an attempt to devise washing techniques related to the chemical characteristics of fibers and detergents.

2. FINDINGS

The initial findings of the research project have identified several ways of alleviating the problems of living in a cramped microgravity environment. The immediate targets are to propose clothing that will help to increase the comfort and efficiency of the astronauts and improve their psychological and physical condition, while not losing sight of aesthetic considerations and the need to ensure garment wearability, bodily thermal stability, hygiene, and the monitoring of biological and physiological functions (although the inclusion of sensors into garments or the fabric itself must in no way prove invasive).

Increasing Comfort and Efficiency.

The absence of gravity is generally viewed as an advantage for movement in space. Once accustomed to microgravity, the body moves quite effortlessly and astronauts are capable of performing "acrobatic maneuvers" with the utmost ease. However, when engaged in physical activity of any kind, the body must be anchored so as to counter the vector force generated by active exercise (Kroemer, Kroemer, & Kroemer-Elbert, 1995). A problem widely encountered during the first days spent in conditions of microgravity is a decrease in motor performance. Because of the absence of gravity and resistance, astronauts finds it takes them longer to carry out all activities. After a few days in –G conditions, astronauts' movements and activities become more accurate, and assessing the amount of physical effort required for an operation no longer appears to be so difficult.

It has been noted that, upon returning to Earth after a long stay on board, astronauts' vertical stability is reduced and there is a loss of muscular strength. They initially tend to drop things, subconsciously believing they will float in the way they do in conditions of microgravity. Such problems tend to disappear after about a week back on Earth. There is no telling as yet how long it will take to recover from the effects of a one-third G environment such as that to be found on Mars (Wichman, Harvey, & Donaldson, 1996).

Microgravity decisively affects not only the human body but also the force-weight of objects, making it more difficult to use and control them. The integrated clothing system is intended to facilitate astronauts' movements and their physical and operative activities by means of a full range of equipment and prostheses built into the garments themselves. These will allow the crew to secure themselves to the internal structure or parts of the garment, and also to carry tools and objects by securing them to the garments themselves. At present, Velcro is the only effective noninvasive system allowing objects to be fixed to the shell of the station itself or to the body of the astronaut, chiefly because it is easy to use and takes up very little space in terms of volume and weight.

Foreseeing New Parameters of Wearability.

Astronauts currently wear the same kind of clothes they wear on Earth. These clothes are obviously unsuited to the peculiar environmental conditions found on board, in part because major changes in body posture and size occur in space. When relaxed, the body takes on a semicurled posture, in which knee and elbow joints are at about 130 degrees, the pelvic angle changes, and the curve of the backbone tends to become flatter in the lumbar and chest region. In addition, the body lengthens by more than 10 centimeters. The head and spine bend forward and the upper limbs bend upward toward the chest by about 45 degrees. This new "neutral posture" that crew members adopt in conditions of microgravity is similar to that adopted by the body under water. It also results from a new-found balance between musculature and the tension of fabrics acting on the different joints (Wichman et al., 1996). Although the "neutral posture" in space does not as such need any countermeasures, apart from alleviating the discomfort that might be felt when bending forward or trying to sit or stand upright, it is necessary carefully to consider the implications for the operator–tool interface. An example is workstation design: Astronauts find it difficult to work at waist height, because they must continuously exert pressure on the arms to keep them at this lower level. Designers must therefore identify which postures and movements are best suited to carrying out specific activities, and take them into account.

The design of space clothing must also take into account the "neutral posture" in space and anticipate new parameters of wearability: weight and tightness, tailoring and stitching of garments, grip and elasticity in relation to postures, and movements typical in microgravity. In addition, the design of the clothes needs to ensure that excess material is not likely to get caught on structural elements inside the ISS.

The Fundamental Role Played by Aesthetics.

In order to improve the level of well-being, the quality of life on board, and onboard social relations, it is essential to use fabrics, shapes, style, accessories, and colors to diversify garments and offer astronauts a wider choice. The pleasure to be derived from wearing a garment capable of enhancing each astronaut's identity is a very meaningful parameter if we consider the confined conditions in which a mixed-sex, multiethnic crew has to live. The integrated clothing system will thus have to highlight the differences between astronauts while taking into account the new parameters of wearability.

Contrary to expectations regarding the formal and aesthetic choices of the integrated clothing system, astronauts feel the need when in orbit to wear clothes not all that dissimilar from those used during training on Earth. To facilitate the crew's adjustment to life on board and to meet the requirements of the users of the ISS clothing system, the best approach seems to be to re-create a domestic and family environment. The integrated clothing system may therefore consider garments for both ground and space activities, by differentiating wearability with respect to comfort, aesthetics, typologies, color, and accessories; thermo-regulation; and antiperspirant, antibacterial, deodorant, and antistatic treatment for fabrics.

Ensuring Thermal Stability and Bodily Hygiene.

Temperatures under normal conditions inside the ISS range from 18.3°C to 26.6°C, but under emergency conditions can range from 15.6°C to 29.4°C. It is therefore imperative for garments to be capable of countering external changes in temperature by keeping body temperature constant.

Confined environments and the shortage of renewable resources, such as water and air, inevitably affect hygiene. Limiting the adverse effects of excessive perspiration, above all when engaged in physical activity, proves to be a considerable advantage, and especially effective when applied to one's "second skin." Thermal-regulating fabrics, interwoven with small silver particles, can absorb excess heat and release it when body temperature falls. The fibers are also capable of curbing the noxious effects of perspiration by transferring excess sweat to the outside, and reducing bacterial proliferation and body odor. This also contributes to improving the quality of air on board.

Speaking of air quality, fabrics and garments should help reduce skin scaling and retain flaked skin particles. This will help reduce an important source of biological pollution inside the ISS, which could even impair the functioning of equipment.

Controlling the Crew's Vital Biological and Physiological Functions.
The monitoring of body values such as temperature, heartbeat, blood flow, breathing, physical stress, and fatigue is at present carried out with equipment calling for the application of sensors for lengthy periods. These restrict astronauts' movements and interfere with the carrying out of assigned tasks and various activities. One area of the integrated clothing project involves specialized sensors that would record astronauts' biosignals without interfering either with the individuals or with their actions. The sensors need to be close to the skin, so they might be incorporated into garments or even into underwear in such a way that the wearer would not be aware of them.

3. CONCLUSIONS

The aim of the VEST research project is to help improve living and working conditions in space. By focusing on the interaction between humans and technology, we can identify systems and tools capable of amplifying human capacities, facilitating efficiency and performance, and increasing the crew's comfort and well-being. The role of industrial design has grown in relation to the quantitative and qualitative complexity of the problems encountered when designing for space. Product systems developed by existing design and clothing companies such as the Benetton Group (clothing for astronauts), Usag (portable and wearable containers for tools), and TechnoGym (light-weight equipment for gymnastics) are examples of how the "strategist designer" comes to act as an intermediary between the aerospace industry and private corporate industries.

In developing product systems for a highly specialized sector such as aerospace, the space industrial designer is confronted with extreme situations. These solutions may be applied to terrestrial problems as well. For example, a clothing system designed for confined environments in conditions of microgravity and utilizing antibacterial processing and thermo-regulating treatment could be very useful during sports competitions, or whenever somebody engages in physical exercise (Dominoni, 1998d).

Given such scope, the task of industrial design is now to identify specific fields of intervention, to create its own methodological and operational tools, and to make its particular competence available in order to solve the problems of designing for confined environments in conditions of microgravity.

REFERENCES AND BIBLIOGRAPHY

Antonutto, G., Capelli, C., & di Prampero, P. E. (1991). Pedalling in space as a countermeasure to microgravity deconditioning. *Microgravity Q, 1*(2), 93–101.

Bandini Buti, L. (1998). *Ergonomia e progetto dell'utile e del piacevole* [Ergonomics and project of the useful and pleasant]. Rimini: Maggioli Editore.

Bluth, B. J. (1985). The social psychology of space travel. In J. E. Katz (Ed.), *People in space*. New Brunswick, NJ: Transaction Books.

Boeing. (1997). Flight crew support and integration (FCS&I). *Space station operations data book (SSODB)*, December.

Bonfantini, M. A., & Zingale, S. (Eds.). (1999). *Segni sui corpi e sugli oggetti* [Signs on the body and on objects]. Bergamo: Moretti & Vitali.

Chiapponi, M. (1999). *Dalla cura delle cose alla cura delle persone. Disegno Industriale e Sanità* [From looking after things to looking after people. Industrial design and health]. Milan: Silvana Editoriale.

Clearwater, Y. (1990a). Auditory environment. In *Space habitability, Avignon Working Days Report*. Exploratory Studies Programme for the Future European Manned Space Infrastructure (EMSI), Atti del convegno.

Clearwater, Y. (1990b). Privacy. In *Space habitability, Avignon Working Days Report*. Exploratory Studies Programme for the Future European Manned Space Infrastructure (EMSI), Atti del convegno.

Columbus team. (1999). Alenia Space Division, *APM IVA Tool List Maintenance*, Technical note, September 29.

Di Bernardo, G. (1997). *Vedo la terra azzurra* [I see the earth blue]. Rome: Editalia.

Dominoni, A. (1998a). L'era spaziale del prossimo millennio [The Space Age in the next millennium]. *OFX*, no. 4, July–August.

Dominoni, A. (1998b). Spin-off: I trasferimenti di tecnologie [Spin off: Transferring technology]. *OFX*, no. 5, September–October.

Dominoni, A. (1998c). Domotica: Lo stato dell'arte [Domotics: The state of the art]. *OFX*, no. 6, November–December.

Dominoni, A. (1998d). IVA clothing support system. In *Explospace*. Workshop on Space Exploration and Resources Exploitation, Atti del convegno, ESA, Cagliari, Sardinia, October 20–22.

Dominoni, A. (1999a). Design dello spazio [Space design]. *Box*, no.1, January–February.

Dominoni, A. (1999b). Habitability issues and lay-out approaches. In *Study of the survivability and adaptation of humans to long-duration interplanetary and planetary environments*. Turin, Italy: Alenia Spazio, November.

Dominoni, A. (2000a). Project research: When the research is inherent to the project. In *Design plus research*. Atti del convegno, Politecnico di Milano, May 18–20.

Dominoni, A. (2000b). *Disegno industriale per la progettazione spaziale* [Industrial design for space planning]. Dottorato di ricerca in disegno industriale, XII ciclo, DI.Tec, Politecnico di Milano, Dipartimento di Disegno Industriale e di Tecnologia dell'architettura [Doctoral research thesis in industrial design, 12th cycle, Polytechnical of Milan, Department of Industrial Design and Architectural Technology], October.

Dominoni, A. (2001). Design spaziale [Spatial design]. *Techne World Wide Magazine*, no. 6.

EUROMIR 95 Daily Activity Plan. (1995). Alenia Spazio internal document.

Gazenko, O. G., Shumakov, V. I., Kakurin, L. I., Katcov, V. E., Chestukhin, V. V., Nikolayenko, E. M., Gvozdev, S. V., Rumyantsev, V. V., & Vasilyev, V. K. (1982). Effects of various countermeasures against the adverse effects of weightlessness on central circulation in the healthy man. *Aviation Space and Environmental Medicine, 53*(6), 523–530.

Green, W. S., & Jordan, P. W. (1999). *Human factors in product design*. London: Taylor & Francis.

Griffin, B. N. (1978). *Design guide: The influence of zero-G and acceleration on the human factors of spacecraft design*. NASA-STD-3000, vol. 1 / rev. B.

Katz, J. E. (Ed.). (1985). *People in space*. New Brunswick, NJ: Transaction Books.

Kirsch, K. (1990). Physical and mental fitness. In *Human space applied psychology. Avignon Working Days Report*. Exploratory Studies Programme for the Future European Manned Space Infrastructure (EMSI), Atti del convegno.

Kirwan, B., & Ainsworth, L. K. (1992). *A guide to task analysis*. London: Taylor & Francis.

Kroemer, K., Kroemer, H., & Kroemer-Elbert, K., (1995). *Ergonomics: How to design for ease and efficiency*. Englewood Cliffs, NJ: Prentice-Hall.

Maldonado, T. (1976). *Disegno industriale: Un riesame* [Industrial design. A re-examination]. Milan: Feltrinelli.

Maldonado, T. (1988). *Il futuro della modernità* [The future of modernity] (4th ed.). Milan: Feltrinelli.

Maldonado, T. (1999). *"Tecnica e società: I nuovi scenari,"* *Il contributo del disegno industriale* [Technics and society: New scenarios, the contribution of industrial design]. Milan: All'insegna del pesce d'oro.

Malerba, F. (1993). *La vetta* [The peak]. Genoa, Italy: Tormena Editore.

Moore, D., Bie, P., & Oser, H. (1996). *Biological and medical research in space*. Berlin: Springer-Verlag.

Murrel, K. F. H. (1965). *Ergonomics. Man in his working environment*. London: Chapman & Hall.

Musso, G., Gaia, E., Battocchio, L., & Colford, N. (1996). *Space hardware for long duration flights*. Tool Box and Elite-S, 47th International Astronautical Congress, Beijing, China.

NASA. (1989). Wardrobe for space. *NASA Information Summaries*.

Rabardel, P. (1997). Gli strumenti dell'uomo. Dal progetto all'uso [The tools of man, from planning to use]. *Ergonomia, 7*, 16–24.

Rambaut, P. (1987). The prevention of adverse physiological change in space station crew members. *Acta Astronautica, 17*(2), 199–202.

Stanton, N. (1998). *Human factors in consumer products*. London: Taylor & Francis.

Wichman, H., Harvey, A., & Donaldson, S. (1996). Remote ergonomic research in space: SpaceLab findings and a proposal. *Aviation, Space, and Environmental Medicine, 67*(2), 151–163.

CHAPTER TWENTY-FOUR

The Technologies
of Body Visualization

Paolo Gerundini and Massimo Castellani

Images of the body play a pivotal role in routine clinical practice today, not only in the diagnosis of many diseases but also in the follow-up and treatment of patients. In this chapter we present a review of the most common imaging methods used in radiology and nuclear medicine.

Soon after their discovery in 1895, X rays started to be used by physicians as a noninvasive means to look inside the human body. Only 6 months after Roentgen's discovery, X rays were being used by physicians to locate bullets in soldiers wounded during the Italian Ethiopian campaign (Blaufox, 1996).

It is now known that X rays consist of electromagnetic radiation, which is generated by the rapid decrease of kinetic energy of electrons pulled towards a tungsten anode in a heated cathode tube. The radiological image is generated through darkening of a film by the radiation transmitted through the patient's body.

Another type of electromagnetic radiation (later called radioactivity by M. Curie) was discovered in 1896 when Henri Becquerel determined the characteristics of certain materials, such as uranium, that spontaneously emit radiation. Today we know that this radiation, originating directly from the nucleus, consists of gamma rays and that an image may be created by their detection with special devices. This process, called radioactivity, is considered the basis of nuclear medicine imaging.

1. CONVENTIONAL RADIOLOGY

Conventional radiology provides images of objects in a two-dimensional (2D) plane. Although it was the first method of imaging to be developed, it is still the most commonly employed type of radiology. The films created by X rays show different features of the body in various shades of gray. The gray is darkest in those areas that do not absorb X rays, whereas lighter shades are seen in dense areas, such as bony structures, that absorb more radiation. Some organs in the body, particularly the bowel, may be visualized by introducing a contrast medium that absorbs the X rays more than the surrounding tissue.

With conventional radiology, both static and dynamic images may be obtained to detect morpho-functional changes in the organs studied. The most frequently performed examinations are skeletal and chest radiography and intravenous pyelography to produce detailed images of the urinary tract from the kidneys to the bladder. Such X-ray examinations are simple, safe, and cost-effective but lack sensitivity and specificity.

2. COMPUTED TOMOGRAPHY

Computed tomography (CT) is a diagnostic procedure in which images are obtained from different angles by X rays passing through the body. The clinical use of CT commenced in the 1970s. Whereas 30 years ago only transverse sections of the body could be obtained, now coronal and sagittal images can be provided. Unlike conventional X-ray examination, CT does not use film to project the transmitted radiation. The incident X rays are collected by a solid-state device or a crystal scintillator, which converts photons into light and subsequently into an electric current in a photoelectron multiplier tube. The electric current has a magnitude that is proportional to the photon energy. The images are processed by a computer, and the CT slices are displayed on a screen and recorded permanently on film or on magnetic tape or optical disks (Ter-Pogossian, 1977).

CT offers certain advantages over conventional radiology because its high resolution allows the detection of smaller lesions (5 mm); moreover, by depicting the shape and borders of organs, soft tissues, and bones, the technique makes it possible to separate overlapping structures. CT images may therefore be employed for the detection and characterization of any mass in the body, particularly in the staging of cancer where the discovery of a small lesion and good definition of anatomic borders may change the therapeutic strategy and prognosis. Furthermore, the spatial resolution of CT images may provide a guide for biopsy when cytological assessment of a mass is required.

New CT equipment (spiral CT scanners) has been developed whose main advantages consist of a shorter image acquisition time and a decrease in the absorbed radiation dose owing to rapid continuous scanning. Moreover, advances in software technology have made it possible to obtain virtual endoscopy of the trachea, bronchi, and esophagus from reconstructed lung CT images (Burke, Vining, McGuirt, Postma, & Browne, 2000).

3. ULTRASONOGRAPHY

After World War II the technique of receiving echoes from a sound transmitted through water (sonar) was investigated for medical use. However, it was only in the 1970s that ultrasonography was introduced as a routine diagnostic imaging method. Because ultrasound examinations are noninvasive and generally considered safe (nonionizing radiation is used), they have become the second most widely used diagnostic imaging modality (Sweet & Arneil, 1975).

A transducer located in a probe converts electrical energy into ultrasound (piezoelectric effect). When the ultrasound beam penetrates the body, it is reflected from tissue interfaces, generating different patterns of echoes. The pulse reflected is reconverted into electric energy by the same transducer.

Because fluid-filled cysts do not reflect the pulse and solid masses have many interfaces, which cause different echo patterns, ultrasonography has always been considered the technique of choice for establishing a differential diagnosis between solid and fluid masses (e.g., in the detection of gallbladder stones or renal cysts). It is also used for exploring the heart (echocardiography).

More recently, ultrasonography has been coupled with Doppler to detect and quantify flow through the heart or carotid arteries (Kremkau, 1991). Today, the use of an ultrasound contrast agent (microbubbles) has been proposed in order to increase the signal-to-noise ratio of images and to quantify coronary flow when they are used in echocardiography (Meerbaum, 1997).

4. MAGNETIC RESONANCE IMAGING (MRI)

The resonance phenomenon consists of the absorption and reemission of radio frequencies by hydrogen, carbon, phosphorus, and fluorine nuclei placed in a magnetic field. This physical phenomenon was discovered in the 1930s and has been called "nuclear magnetic resonance." The length of time these response signals are emitted after an atom has been stimulated by radio waves is called "relaxation time" (T1 and T2); it may vary widely, depending on the tissue components being examined (Rosen & Brady, 1983).

When a patient is being scanned with magnetic resonance (MR), the response signals emitted by the nuclei are revealed by an antenna and forwarded to a computer for processing. MRI (magnetic resonance imaging) data acquisition is three-dimensional (3D) in origin and sagittal, coronal, and transaxial images can be displayed at present.

Because the relaxation time reflects the chemical composition of tissues, MRI may provide not only anatomic images but also an accurate characterization of a mass composition (Pykett, 1983). In this regard, the injection of a paramagnetic contrast agent, for example gadolinium, may improve the visualization of certain body structures by altering the relaxation times of the stimulated nuclei and also the image contrast.

5. NUCLEAR MEDICINE IMAGING

In conventional X-ray or CT scan procedures, the electromagnetic radiation originates outside the patient and the images are projected onto the film directly from the radiation, attenuated by the different structures of the body or by means of computer elaboration. In nuclear medicine, a radioactive tracer is generally injected into the patient—although it may also be introduced by inhalation or swallowing—and the gamma rays emitted by the source inside the patient are revealed by a detector called a "gamma camera." After photon collection by a crystal scintillator, the processes are similar to those observed with CT, and the light signals are converted into an electronic pulse (Budinger, 1977).

Because the tracer compounds are usually substances that enter into physiological processes, the nuclear medicine image reflects the function of the organ rather than its anatomy. As a consequence, different tracers can be used to study nearly all the organs of the body depending on their physiology. For example, a bone scan detects local changes in bone metabolism in response to tumors, fractures, and other abnormalities. These changes can often be seen before any changes are manifest on bone radiographs (about 40 percent of the bone mass may be lost without osteolysis appearing on conventional radiographs), so bone scintigraphy is considered the most sensitive imaging technique for the detection of bone disease.

Before the 1970s only 2D planar images could be obtained. With the advent of rotating gamma cameras, transaxial, coronal, and sagittal images became available. This method of acquisition has been called "single photon emission tomography" (SPET). Finally, 3D volumetric image of the organ may be obtained by combining series of 2D projection images to create the third dimension (Kirsch et al., 1981).

Tomographic images have a better resolution than planar images and better diagnostic sensitivity. 3D emission computed tomography provides a more accurate localization of the physiological or pathological distribution of the tracer injected.

6. POSITRON EMISSION TOMOGRAPHY (PET)

Like CT, MRI, and SPET, positron emission tomography (PET) provides tomographic images. Unlike planar and SPET images, which are obtained by means of a single photon, in PET the annihilation of a positron emitted from the nucleus and an electron from its surroundings generates two photons emitted at a 180° angle. These gamma rays are then revealed within a coincidence time window by a ring of scintillator detectors. Although 3D images are also generated with this technique, the sensitivity in the detection of lesions and the resolution are better than those of SPET (Bendriem & Townsend, 1998). Because the tracers used are physiological components of organic matter (carbon, nitrogen, oxygen, and fluorine), pathological changes in the metabolism of amino acids, glucose, or ammonia can be studied in the body.

One of the most important applications of PET is imaging with 18-Fluorodeoxyglucose (18FDG). FDG-PET is currently used in neurology, where it plays an important role in the differential diagnosis between recurrence of brain tumors and radionecrosis. This application is based on the concept that tumor tissue is able to consume glucose, whereas necrotic tissue is not (Deshmuck et al., 1996). Another example is the typical distribution of FDG in the brains of patients affected by Alzheimer-type dementia. This pattern may contribute to the early diagnosis of this disease (Mielke et al., 1994).

In oncology, FDG-PET plays an increasingly important role in the detection and staging of many types of cancer, as well as in the assessment of the efficacy of chemotherapy (Braams, Pruim, & Freling, 1995; Lindholm, Leskinen, & Nagren, 1995). In cardiology, information about the coronary reserve and residual viability of the myocardium after infarction may be provided by ^{13}N-ammonia and FDG-PET, respectively (Marshall et al., 1992).

7. NEW PERSPECTIVES OF BODY IMAGING

Medical imaging must be able to provide not only a diagnosis but also information about the physiopathology of disease; it is important to know not only whether the patient has a disease but also how it can be managed. Moreover, the physician should have a means to assess the efficacy of treatment and to ascertain whether or not a particular treatment will impair other body functions.

In oncology, the metabolism of a lesion observed with CT may be characterized by PET, and both the prognosis and the response of a neoplastic mass to chemotherapy may be assayed by a decrease in glucose uptake. In neurology, PET activation images of the brain obtained with ^{16}O-labeled water allow the surgeon to establish which areas are at risk of injury during neurosurgery (Petersen et al., 1988; Spetsieris, Moeller, & Dhawan, 1995).

In conclusion, at the beginning of the third millennium, the integration of multiple imaging modalities is considered mandatory for clinical routine. The merging of functional SPET or PET images with anatomic MR or CT images (Pellizzari et al., 1990) is an important challenge for the future.

REFERENCES AND BIBLIOGRAPHY

Bendriem, B., & Townsend, D. W. (Eds.). (1998). *The theory and practice of 3D PET.* Dordrecht, Netherlands: Kluwer Academic.

Blaufox, M. D. (1996). One hundred years of radioactivity (1896–1996). *Seminars in Nuclear Medicine, 25,* 145–154.

Braams, J. W., Pruim, J., & Freling, N. J. M. (1995). Detection of lymph node metastases of squamous-cell cancer of the head and neck with FDG-PET and MRI. *Journal of Nuclear Medicine, 36,* 211–216.

Budinger, T. F. (1977). Instrumentation trends in nuclear medicine. *Seminars in Nuclear Medicine, 7,* 285–297.

Burke, A. J., Vining, D. J., McGuirt, W. F., Jr., Postma, G., & Browne, J. D. (2000). Evaluation of airway obstruction using virtual endoscopy. *Laryngoscope, 110,* 23–29.

Deshmuck, A., Scott, J. A., Palmer, E. L. et al. (1996). Impact of fluorodeoxyglucose positron emission tomography on the clinical management of patients with glioma. *Clinical Nuclear Medicine, 21,* 720–725.

Kirsch, C. M., Moore, S. C., Zimmerman, R. E. et al. (1981). Characteristic of a scanning, multidetector, single-photon ECT body imager. *Journal of Nuclear Medicine, 22,* 726–731.

Kremkau, F. W. (1991). Principles of color flow imaging. *Journal of Vascular Technology, 15,* 104.

Lindholm, P., Leskinen, S., & Nagren, K. (1995). Carbon-11-Methionine PET imaging of malignant melanoma. *Journal of Nuclear Medicine, 36,* 1806–1810.

Marshall, R. C., Tillisch, J. H., Phelps, M. E. et al. (1992). Identification and differentiation of resting myocardial ischemia and infarction in man with positron emission computed tomography. 18 F-labeled fluorodeoxyglucose and N-13 ammonia. *Circulation, 85,* 1347–1353.

Meerbaum, S. (1997). Microbubble fluid dynamics of echocontrast. In N. C. Nanda, R. Schlief, & B. B. Goldberg (Eds.), *Advances in echo imaging using contrast enhancement* (pp. 11–37). Dordrecht, Netherlands: Kluwer Academic.

Mielke, R., Pietrzyk, U., Jacobs, A., et al. (1994). HMPAO SPECT and FDG PET in Alzheimer's disease and vascular dementia: Comparison of perfusion and metabolic pattern. *European Journal of Nuclear Medicine, 21,* 1052–1060.

Pellizzari, C. A., Chen, G. T. Y., Spelbring, D. R. et al. (1990). Accurate three dimensional registration of CT, PET, and / or MR images of the brain. *Journal of Computer Assisted Tomography, 13,* 573–579.

Petersen, S. E., Fox, P. T., Posner, M. I. et al. (1988). Positron emission tomography studies of the cortical anatomy of single-word processing. *Nature, 331,* 585–589.

Pykett, I. L. (1983). Instrumentation for nuclear magnetic resonance imaging. *Seminars in Nuclear Medicine, 13,* 319–328.

Rosen, B. R., & Brady, T. J. (1983). Principles of nuclear magnetic resonance for medical application. *Seminars in Nuclear Medicine, 13,* 308–318.

Spetsieris, P. G., Moeller, J. R., & Dhawan, V. (1995). Visualizing the evolution of abnormal metabolic networks in the brain using PET. *Computerized Medical Imaging and Graphics, 19,* 295–306.

Sweet, E. M., & Arneil, G. C. (1975). An introduction to the use of diagnostic ultrasound. *Seminars in Nuclear Medicine, 4,* 289–298.

Ter-Pogossian, M. M. (1977). Basic principles of computed axial tomography. *Seminars in Nuclear Medicine, 7,* 109–127.

CHAPTER TWENTY-FIVE

Conclusion:
Bodies Mediating the Future

Leopoldina Fortunati, James E. Katz, and Raimonda Riccini

The contributors to this volume have examined the phenomenological consequences of the diverse social forces that are increasingly converging on the human
body. These processes can be very different from one another in their origin and
intention, but share the same focus: the human body. As our authors have detailed, these forces can be seen as being driven strongly by progress in technology
and the resulting advances in the ways and means of communication.

This extraordinary convergence also raises many significant problems for
those who wish to steer developments in a direction that advances the interests of
some individuals and groups at the expense of others. What we wish to argue here
is that it is more useful to consider the body as both an inherent unity in itself as
well as a linking element between the essentially social human and the larger
physical and cultural environment. As reviewed in this volume, there is a strong
tendency to separate the body analytically into its various manifestations. Each intellectual discipline likewise views the body from its own perspective, including
the biological, physiological, and psychological aspects. These include perspectives on its social and religious history, its covering, fashion, and adornment, and
its place in artistic and literary endeavors. In particular, the fashion and the consumer perspective includes the constructed body (fashion, but also sport, diet, and
image), the "artificial" body (plastic surgery, psychoactive drugs, and assisted pro-
creation), and the extended dimensions of the body (from communicative prostheses to medical ones such as artificial limbs and joints and pacemakers).

As sociologist Chris Shilling (1993) noted, social theories of the human body
have progressed mainly through these segmented areas. The numerous points of
view represented here on the theme of the body and its future suggest, as did Shilling, that new aggregations of formerly separate spheres are necessary for further
progress to be made. Certainly several authors in this volume have indicated areas
of possible cooperation.

Although a crude lumping together of concepts runs the risk of misleading and irritating readers more than informing and delighting them, we none the less offer a rather large clump. We argue that the main battleground between the forces of culture and technology is becoming the human body (Chernin, 1982; Connel, 1987; Frank, 1991, 1995; Freund & McGuire, 1998; O'Brien, 1981, 1989; Orbach, 1988). If this view is endorsed by other social scientists, this insight may provide a useful lens through which to investigate this intriguing nexus. In fact this view shows how the body continually abolishes the border between nature and technology by converting one into the other (Liuccio Tafuri, 2001). At the same time, it may allow us to go beyond "naturalistic" and "constructionist" visions of the body, which in themselves are both reductive because the body in its unity is simultaneously a biological and a social phenomenon (De Nardis, 1999; Melucci, 1991).

The human body we inhabit is smart in many ways. It can do things by itself that are beyond the capability of the world's most advanced scientific and engineering laboratories. It can create red blood, precisely distinguish between "red" and "read," and "see red" when insulted. The ever-mysterious body provides us with a full sensory experience, and yet separates us from the world so that we can never have direct contact with it. We share existence with the body, but the boundary of this sharing is evanescent and recedes even as we seek to probe and inspect it. The body is different from who we are, but we are generally indistinguishable from it. The body in part executes our will with near-magic precision, and in part capriciously flouts us. With intriguing ambiguity, it is partly under our control and partly in control of us. It is who we are and makes us what we are; and we, to an expanding degree, make it what it is (Schutz, 1940/1986). This explains the enormous categorial elasticity of the concept of body.

At present we cannot do without the body, although, as shown in some chapters of this volume, individuals as well as research teams are exploring ways to dispose of, or at least mitigate, this physical baggage. However, whereas for some it is a state of being from which escape is sought, others pursue a more intense or seamless integration with it.

The struggle to enforce a given resulting explanatory framework has been an important component in the world histories of religion, magic, and superstition. It has also been an important aspect of political and social movements, and even of military conflict (Day, 2002).

In recent times, computer engineers, psychologists, sociologists, cognitive scientists, postmodern critics, and even industrial design experts have all weighed in on the subject of the body. Our purpose in editing the chapters in this volume is to present the state of the art of the debate and encourage further intellectual development of the area. We seek to offer multiple lenses from, as it were, the same platform. Each chapter serves as a lens through which we can perceive some aspects of the body in a social context. However, each lens has its own field of focus. It can do much to bring clarity and insight to the elements within its focal field; but it inevitably blurs or misses important aspects of the problem in the conceptual foreground and background. This is a sad truth, in physics no less than in metaphysics (Bronowski, 1976). Nevertheless, we believe a special strength of this volume is the emphasis our authors bring to bear, not the subject. By juxtaposing the details and insights that these respective lenses reveal, we hope to achieve a more comprehensive view of the body among mediating forces than is possible from any one perspective. Perhaps a good analogy is the multiple interlocking mosaic fragments which, entomologists claim, combine to give the house fly an effective and broad picture of the world.

Certainly the body has been an object of contemplation and study for millennia, as the rude stone figures of prehistoric humans and the magnificent arching movements of the figures of the cave paintings reveal, every bit as cunningly as the remastered photos of faces that peer out from magazine covers at newsstands. In terms of intellectual glass bead game((Hesse, 1969), there has been much recent attention to the question of the body (Bartlett, 1999). Yet the integration of body in other areas, such as the phenomenology of robots, exploration as a literary theme, use in psychoanalytic precepts, and the use of technology as extensions of the body, has not been systematically explored. Although we do not claim that we have a system, many contributors to this volume provide new understanding of these areas.

The wide variety of novel perspectives in this volume reveal the body as a socially constructed (but physically real, though not easily physically delimited) element. These perspectives are all connected by the themes of communication and ICTs (information and communication technologies), such as health and well-being, style and functionality, and fashion as self-expression, social control, and group affirmation. Many of our contributors describe current conditions or investigate and interpret the past. Others indulge in speculation about the future. As editors we have set them to explore the body as it floats in hermetic containers zooming about the vacuum of outer space, or sits in the thin steel containers of automobiles that career down highways at seemingly galactic speeds. We have invited them to consider what goes on just above the skin's surface as well as actually on or under the skin itself. We wanted the internal aspects of the body examined as well as its social environment. Our contributors espy the body in the manner in which it is clothed and the way in which it is marked, disfigured, and reconfigured.

From the perspective of the contributors to this book, the body is also seen as a tantalizing way for people to express themselves as who they are, what they have been, and who they would like to be. In the context of our book's themes, they have presented studies that analyze the way the body is extended and distorted by technology for the sake of communication and fashion. This was shown in the studies of the mobile phone, cars, clothes, and skin. Likewise we have seen how social demands, reflected in developments in technology and fashion, have projected the body into outer space and covered it with exotic fabrics that seemingly have a life of their own.

Fashion and technology both have important elements of communication in their design and deployment. Fashion is predicated on increasingly sophisticated technology, and technology is applied to the body with increasing sophistication. Our contributors have shown this with force, formal logic, and an array of data, evidence, and anecdotes from various sources. Their chapters have demonstrated how these issues permeate literature, gender studies, and social interaction. The body might at a simple level be seen as a machine, as argued by de la Mettrie (1748/1996). Yet few would disagree that to become one with a body can be a sublime and powerful (and painful and frustrating) experience. Moreover, the body is the physical embodiment of the human spirit and psyche. As such, it is also a profound form of communication. As the adage in the sociology of communication has it, we cannot *not* communicate." In the case of the body, the medium often is the message. And the appearance of the messenger contributes to enunciating the message.

However, we also think that a parallel perspective could yield insights of value. In particular, we have developed this perspective in another project, which we describe as the "Machines That Become Us" perspective and which will be published in the book of the same title (Katz, 2003). This perspective goes in the opposite direction to that proposed by MIT computer scientists, who believe we will come to accept robots and other machines as full emotional and legal equals. This is the conclusion of a long line of studies that, following thinkers such as Bruno Latour, judge technological objects to be so aggressive and numerous as to have themselves become a new population of social "subjects" or "technological bodies" (see Landowski & Marrone, 2002). As we have seen again and again, this process takes place not only through the progressive artificialization of the human body, but also through a "personalization" of objects, such as to lead to the definitive loss of the distinction between people and things (including machines). Our point of view is that, even though people may "become" machines, in the sense that machines will be increasingly integrated into the bodies, lives, and social environments of people, machines will never replace, let alone become coequals with, people on the level of emotion, feeling, justice, or rights.

REFERENCES AND BIBLIOGRAPHY

Bartlett, D. (Ed.). (1999). *Body in transition*. Zagreb, Croatia: University of Zagreb.

Bronowski, J. (1976). *The ascent of man*. Boston: Little, Brown.

Chernin, K. (1982). *The obsession: The tyranny of slenderness*. New York: Harper & Row.

Connel, R. (1987). *Gender and power*. Cambridge, England: Polity Press.

Day, R. (2002). *The modern invention of information: Discourse history and power*. Carbondale: Southern Illinois University Press.

de la Mettrie, J. O. (1996). *Machine man and other writings* (Cambridge Texts in the History of Philosophy). Cambridge, England: Cambridge University Press. (Original work published 1748)

De Nardis, P. (1999). *Sociologia del limite* [Sociology of limits]. Rome: Meltemi.

Frank, A. W. (1991). *At the will of the body*. Chicago: University of Chicago Press.

Frank, A. W. (1995). *The wounded storyteller: Body, illness and ethics*. Chicago: University of Chicago Press.

Freund, P., & McGuire, M. (1998). *Health, illness and the social body*. Englewood Cliffs, NJ: Prentice-Hall.

Hesse, H. (1969). *Glass bead game: Magister ludi* (R. Winston & C. Winston, Trans.). New York: Holt.

Katz, J. E. (2003). *Machines that become us: The social context of personal communication technology*. New Brunswick, NJ: Transaction.

Landowski, E., & Marrone, G. (Eds.). (2002). *La società degli oggetti. Problemi di interoggettività* [The sociology of objects: Problems of inter-objectivity]. Rome: Meltemi.

Liuccio Tafuri, M. (2001). La sociologia e il corpo: L'interpretazione naturalistica [The sociology of the body: The naturalistic interpretation]. *Sociologia, 35*(3), 63–70.

Melucci, A. (1991). *Il gioco dell'io* [The game of the self]. Milan: Feltrinelli.

O'Brien, M. (1981). *Reproducing the world*. Boulder, CO: Westview Press.

O'Brien, M. (1989). *The politics of reproduction*. London: Routledge & Kegan Paul.

Orbach, S. (1988). *Fat is a feminist issue*. London: Arrow Books.

Schutz, A. (1986). *The phenomenology of the social world* (G. Walsh & F. Lehnert, Trans.). Evanston, IL: Northwestern University Press. (Original work published 1940)

Shilling, C. (1993). *The body and the social theory*. London: Sage.

Author Index

Note: *n* indicates footnote

A

Aakhus, M., 6, *11*
Aakhus, M. A., 75, 76, 77, 86, *86*
Abraham, K., 140, *145*
Ackerman, D., 20, *21*
Adorno, T. W., 47, *49*
Ainsworth, L. K., *208*
Alberoni, F., *146*
Alberti, B., 55, *58*
Albini, A., *173*
Albinson, C., 158, *160*
Alfano Miglietti, F., *49*
Allen, B., *121*
Anceschi, G., 177, *184*
Andrieu, B., *145*
Antinucci, F., 178, *184*
Antonutto, G., 202, *207*
Appleton, J., 196, *199*
Argyle, M., 125, *130*
Arneil, G. C., 211, *214*
Attfield, J., *121*
Azolay, J. F., 156, 158, *160*
B
Babbage, C., 147, *153*
Ballard, J. G., 56, *58*
Bandini Buti, L., *207*
Banta, M., 117, *121*
Barber, B., *146*
Barthes, R., 164, 166, *167*, 178*n*, *184*
Bartlett, D., 217, *218*
Bataille, G., 47, *49*, *145*
Battegay, A., 115, *122*
Battocchio, L., *208*
Baudrillard, J., 63, *70*, 151, *153*
Bell, D., 43, *49*
Belloni, L., 17, *21*
Belotti, E. G., 55, *58*

Bendriem, B., 212, *213*
Benjamin, W., 166, *168*
Berg, A. -J., 120, *121*
Berlo, A., *121*
Bernasconi, S., *145*
Berndt, T. J., 100, *102*
Bernège, P., *121*
Bernieri, F. J., 126, *130*
Bertora, F., 183, *184*
Bettinelli, E., *121*
Bie, P., *208*
Blanchot, M., 47, *49*
Blaufox, M. D., 209, *213*
Blumer, H., 155, *160*
Bluth, B. J., 202, *207*
Bocchi, G., 198, *199*
Bodei, R., 42, *49*
Bogo, D., 148, *153*
Bolen, J. S., 4, *10*
Bonfantini, M. A., *207*
Borden, L., *122*
Boreano, P. L., 114, *121*
Borel, F., *145*
Borrows, R., *146*
Bose, C. E., 114, *121*
Boultwood, A., 155, *160*
Bovone, L., *145*
Bowlby, J., 128, *130*
Braams, J. W., 213, *213*
Bradbury, T. N., 128, *131*
Braddock, S. E., 147, 148, *153*, 170, *173*
Brady, T. J., 211, *214*
Brand, S., *145*
Brega, G. P., 47, *49*
Brohm, J. M., *145*
Bronowski, J., 217, *218*
Brooks, R. A., 1, *10*
Brosnan, M. J., 107, *111*
Browne, J. D., 210, *213*

221

Subject Index

Note: *f* indicates figure, *n* indicates footnote

Printed in the United States
by Baker & Taylor Publisher Services